SCIENCE

50 ESSENTIAL IDEAS

SCIENCE

50 ESSENTIAL IDEAS

ANNE
ROONEY

SIRIUS

This edition published in 2022 by Sirius Publishing, a division of
Arcturus Publishing Limited,
26/27 Bickels Yard, 151–153 Bermondsey Street,
London SE1 3HA

ISBN: 978-1-3988-2066-1
AD008040NT

Printed in China

Contents

Introduction 7
1. Uncuttables 9
2. Into the void 13
3. Pass it on 15
4. Cutting the uncuttables 19
5. Painless surgery 24
6. Elementary 27
7. Imposing chemical order 31
8. All the way down 36
9. Starting with a Bang 42
10. Units of life 46
11. A central star 50
12. In balance 53
13. Energy from atoms 59
14. Rock of ages 63
15. Making planets 67
16. The smallest living things 71
17. Spreading diseases 75
18. Shining a light on light 79
19. From radio to gamma rays 83
20. The origins of life on Earth 87
21. Pushing together, pulling apart 91
22. Starlight star bright 95
23. Dark forces at work 99
24. From crust to core 105
25. Smashing plates 110
26. Heat and gases 114

27. Too hot, too fast 117
28. Atoms lining up 120
29. DNA 124
30. The matter of matter 128
31. Keep it clean 132
32. Kill the germs 136
33. You can't lose 140
34. Everything in order 143
35. Cells within cells 146
36. Vax v. anti-vax 150
37. Holes in space? 154
38. Round and round 158
39. A helping hand for the chemistry of life 162
40. Layer after layer 165
41. Never out of energy 168
42. Non-chaotic chaos 170
43. Survival of the fittest 172
44. Evolution, now with genes 178
45. Dead and gone 181
46. It's all relative 185
47. Smaller than small 188
48. A cat in a box 193
49. Gaia 195
50. Are we alone in the universe? 197
Index 204
Picture Credits 208

Introduction

Science is one of humankind's greatest achievements. It explores how the universe around us (and within us) works, seeking to explain it in clear, logical and consistent ways. An understanding of science is crucial to being an informed, engaged citizen. The 50 ideas outlined in this book will give you a good grounding in the science on which the modern world is built, enabling you to take a critical and balanced view of science stories you see in the media.

We traditionally divide science into different areas, such as biology and physics, but in reality they overlap. Indeed, one of the ways we know science is largely 'right' is because its explanations fit together. Chemistry explains and underlies biology; the physics of atoms explains and underlies chemistry.

ESSENTIAL SCIENCE IDEA #0: THE SCIENTIFIC METHOD

Science itself is the primary scientific idea. It can be traced to Thales, living in Greece in the 6th century BC. He sought to explain natural phenomena in terms of rational, physical causes, without recourse to supernatural entities. The idea was developed more fully by Aristotle. Modern science follows what has become known as the 'scientific method', rooted in the work of Francis Bacon and others in the 17th century. Scientists often work from observations to discover general laws and causes. If the laws suggested are correct, it's possible to make predictions from them that will be borne out. Science can start with a bottom-up approach (deduction), working from specific observations towards general principles, or a top-down approach (induction), working from a suggested principle to see if it's confirmed in specific observations.

A typical process has a scientist asking a question, perhaps as the result of an observation. They formulate a hypothesis, which is a suggested explanation that can be tested by rigorous experiments and observations. This might involve physical trials, or a great deal of mathematical modelling (we can't recreate the universe to test Big Bang theory). The findings of the investigation are compared with the hypothesis, which might then be rejected, refined, or endorsed.

If thorough investigation endorses a hypothesis, science might end up with a new theory. A theory is not an untested idea; it's an intellectual model of how the world is. The theory of evolution is not just a suggestion, or a possible explanation that can be weighed against others. It's a fact, as far as we can tell, borne out by more than a century of investigation. The mechanisms of evolution have been discovered and the theory refined in that time.

The body of scientific knowledge is built on previous discoveries, so something like the modern statement of evolution (the modern evolutionary synthesis) has grown incrementally as scientists over decades have contributed expertise, ideas and explanations. Modern science is very much a co-operative and cumulative enterprise.

EVERYTHING CAN CHANGE

Refinement and restating of theories are essential to science. Science recognizes no immutable truths that can't be challenged; everything is up for grabs. If new evidence appears to overturn an established idea, it is examined and tested. Intransigence is a failing of individuals, not of science itself. A good example is the Michelson–Morley experiment to prove the existence of 'luminiferous aether' in 1887 which proved instead its non-existence, furthering science in a different direction (see page 85).

FIFTY AND COUNTING...

There are, of course, more than 50 important ideas in science, and more will appear. Some new, earth-shattering discovery might soon overturn some of these, or deserve a place on its own merits. Perhaps the discovery of life beyond Earth, proof of the existence of gravitons (quantized particles of gravity), or some revolutionary new means of energy production, or even some great insight into the nature of time. Science is far from over, and this book is just a snapshot.

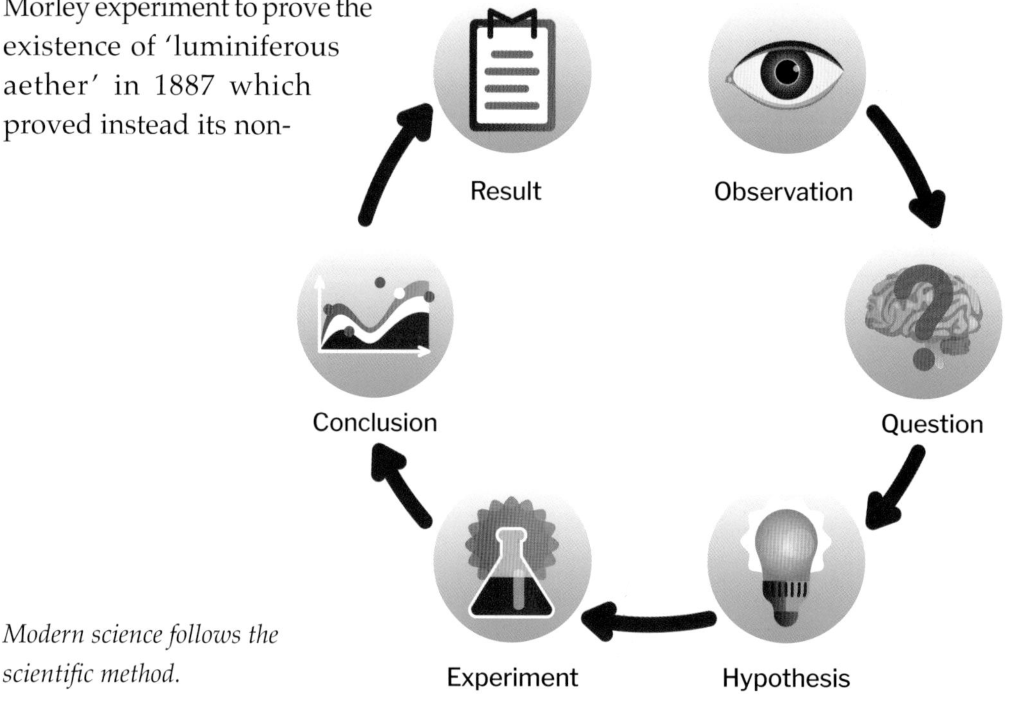

Modern science follows the scientific method.

1

Uncuttables

The atomic theory of matter

If you look at everything surrounding you right now, you'll see a huge variety of different matter – and some you won't see at all, as the air isn't visible. We now have a well-established theory that matter is made of tiny particles called atoms, fully supported with evidence. The Ancient Greeks are the first people known to have thought about the organization of matter.

'UNCUTTABLES'

Our modern word 'atom' comes from the Ancient Greek *atomos*, an adjective that means 'uncuttable'. The philosophers Leucippus and Democritus are credited with introducing the idea 2,500 years ago that matter is made of very tiny particles that can't be further divided. According to Democritus, everything is made of tiny particles that exist in a void. These particles, atoms, can't be created or destroyed, but they can move around. Atoms, he claimed, are of different sizes, shapes and orientations and can

Paradoxes of divisibility

Democritus put a limit on the divisibility of matter with his 'uncuttables', but the idea was also applied elsewhere. The paradoxes of Zeno question whether time and distance can be infinitely divided. Zeno's paradoxes suggest that movement and any other kind of change should be impossible. If an arrow travels, say, 100 m (328 ft) in one second, it would then travel 50 m (164 ft) in half a second, and so on. When it is one metre (3 ft) from its target, it will travel half a metre (20 in.) in half the next hundredth of a second, but still be half a metre away. It will travel a quarter of a metre in half the remaining time, and so on. If time and distance can be infinitely divided, there never comes a point when it has to travel the last tiny distance, so it will never get there. But, of course, it does.

temporarily link to others, providing the variety of different types of matter. As the clusters of atoms break up and the atoms regroup into new clusters, matter changes.

BACK TO BASICS

Atomism next surfaced in 17th-century Europe, particularly in the work of Anglo-Irish chemist Robert Boyle. Boyle regarded atoms as all being alike in respect of their impenetrability, but differing in size, shape and motion.

Robert Boyle was a proponent of atomism in the 17th century.

Size and shape are primary properties of matter. Secondary properties, including colour, smell, elasticity, density and so on, he considered to be produced by the particular combination and arrangement of atoms. Boyle also referred to 'natural minima' or 'least parts', as being the smallest particles of a substance that have the substance's secondary properties. These weren't (usually) the same as atoms, though they might sometimes come close to our concept of molecules. For Boyle, as for the ancients, all atoms were made of the same material, regardless of the type of matter they eventually found themselves in.

This view was shared by Isaac Newton, too. He made atoms subject to his laws of movement, so that between collisions, atoms would observe the laws of momentum and inertia. Newton accounted for the density of matter of different types by how closely packed atoms are in it, but he also allowed that there is a lot of space within atoms, as light can pass through gold leaf even though gold is very dense. But Newton introduced forces to the purely mechanistic atoms of Boyle. Newton's atoms could be held together in matter by forces of attraction, or could dissociate because of repulsion. This idea could explain how one chemical would displace another in a reaction – if there was a greater attraction between A and C than between A and B, A would combine with C and B would be somehow thrown out. This explanation, though, couldn't be used to make predictions: it couldn't tell you what would happen between A and D.

In the 18th century, forces became more important in all areas of science. Ruđer Bošković, from what is now Croatia, went so far as to strip atoms of materiality and make them centres of force – including gravity – so that they could still confer mass on matter.

ATOMS, AT LAST

A connection between atoms and chemistry was successfully made, at last, by the English chemist John Dalton. He suggested that the chemical elements are made up of 'ultimate particles', or atoms, with each element having its own unique design of atom. When Antoine Lavoisier compiled the first modern list of elements (see page 28), he didn't link the idea to atoms.

Dalton used the proportions in which chemicals combine to tell him the relative atomic weights of elements, using hydrogen as the basis (atomic weight 1). He was mistaken when he assumed that elements combined in the ratio

John Dalton first proposed that each chemical element has its own unique design of atom.

Dalton devised symbols for the atoms of 20 elements. Of these, six are now recognized as compounds.

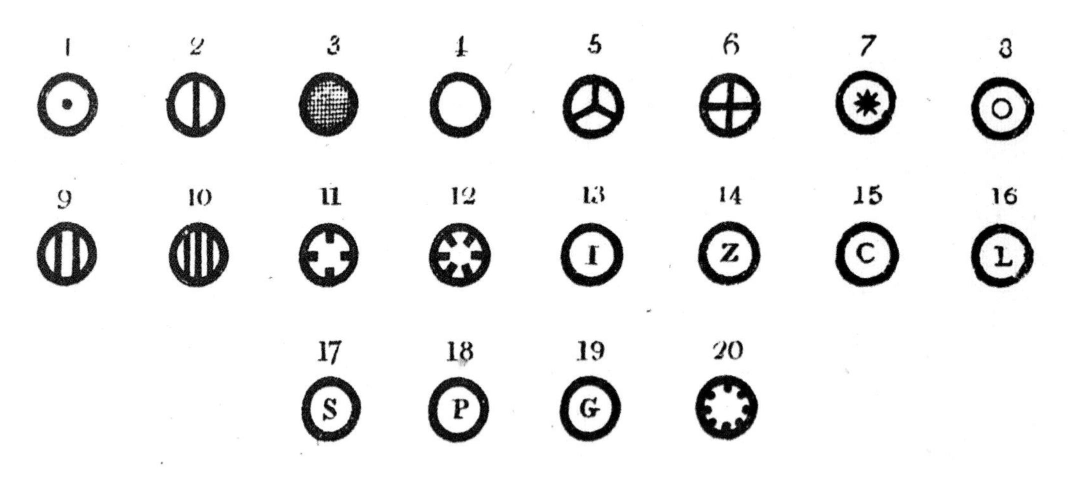

1:1 – so assuming one hydrogen atom combined with one oxygen atom to make water, rather than two hydrogen atoms with one oxygen atom. He dropped the traditional assumption, starting with Democritus, that all atoms are made of the same substance and differ only in properties such as size and mass. In Dalton's system, atoms of different elements were all made of different types of stuff.

Final proof that atoms exist came with Albert Einstein's explanation of an observation made by the botanist Robert Brown in 1827. While examining pollen grains under a microscope, Brown had seen them moving randomly around on their own. It became known as Brownian motion, but its cause was unknown. Brown had been unable to explain it, though he'd seen the same movement with other tiny particles, including some that were not from living things. In 1905, Einstein explained the motion: it's produced by the tiny light particles being jostled by the random movements of water molecules. The molecules are too small to see, but we see the effect of their collisions with pollen grains.

2

Into the void

The existence of a vacuum

We now take for granted that we can create a vacuum – a space containing nothing. But for a long time, the idea of the void was highly contentious.

THE SPACE WITHIN MATTER

The idea that there might be empty space was first debated by the Ancient Greeks. When Democritus described a world made up of atoms, he had them existing in a void, something we would recognize now. But not all thinkers agreed with him. On one side were those, like Parmenides and Aristotle, who claimed that there is no empty space, but matter is entirely continuous. Parmenides argued on this basis that movement and change don't exist. As it's fairly clear that things do move and change, Democritus's view of a world where atoms can move through the void sounds more compelling.

The possibility or impossibility of a void remained a philosophical debate for many centuries. In some cases, it even seemed to be ruled by the meaning of words: nothing can't exist because 'existence' requires the presence of some matter. Eventually science won out over philosophy with a practical demonstration. In 1654, the German scientist Otto von Guericke resolved the question by inventing and demonstrating a pump to create a vacuum. He fitted two metal hemispheres together to make a closed sphere and withdrew all the air from it. In a spectacular public demonstration, two teams of horses were unable to pull the hemispheres apart. The reason is not that the vacuum 'sucks' the halves together, but that the pressure of the air outside the hemispheres pushes them hard together. The horses are trying to pull against air pressure.

As evidence emerged of the existence of atoms, our understanding of the world returned to something like Democritus's void containing particles of matter. Where atoms are more densely packed, the matter they make tends to be a solid. In between solid objects, there could be liquids or gases (or empty space). In liquids and gases, atoms are more widely spaced and move more freely.

OUTER SPACE

Some of the Ancient Greeks accepted that outside Earth there might be empty

space, and most people now think of outer space as a vacuum. It is close to being a perfect vacuum. There is very little matter in the gaps between stars, planets and other bodies. Gravity draws any stray matter towards other clumps of matter, making space very uneven in terms of the density of matter in it. There are very dense concentrations of matter (such as Earth) and then some very empty areas between star systems and between galaxies. The average density of space is just under six protons per cubic metre, but in the cosmic void – the space between crowded areas – it's less than one atom per cubic metre.

Von Guernicke's dramatic demonstration of the Magdeburg hemispheres proved a vacuum is possible.

In quantum mechanics, a vacuum must have no matter particles and also no photons. As space is everywhere, subject to the cosmic microwave background radiation (see page 44), there is no true vacuum in this sense.

3

Pass it on

The principle of heredity

It's obvious by looking around at family, pets, farm animals and even in the garden that one generation of organisms passes on features to the next. Many of us look like our parents, our children and our siblings. As humans, we have exploited inherited characteristics in agriculture for millennia, since long before we had any understanding of how characteristics were passed on.

The earliest farmers bred from their most productive animals and saved seed from their best crop plants, slowly changing the nature of crops and livestock so that now they bear little resemblance to their wild ancestors. The first investigation of patterns of inheritance, though, came fewer than 200 years ago.

THE MONK AND THE PEAS

In a monastery garden in Brno (now in the Czech Republic) in 1857, an Augustinian monk called Gregor Mendel began a careful study of pea plants to investigate the inheritance of characteristics. He focused on seven different features of the plants, including flower colour, seed colour and seed shape. For each characteristic he began by breeding two lines of peas, one with each version of a feature: for instance, white flowers or mauve flowers. Once he had established his true-breeding lines for each feature, he began to cross-breed plants and keep track of the outcomes.

He found that in all seven features, the first generation of crosses (for instance, tall pea × short pea), produced only one form (tall pea, in this case). Breeding from the next generation, he found a mix, but always in the ratio 3:1 (so three tall peas to one short pea). He called the feature that was always seen in the first generation, and was most frequent in the second, the dominant trait. The other he called the recessive trait. He concluded that a pea plant has two heritable 'factors' for each feature. A dominant factor will always mask a recessive factor, so if a plant inherits one tall (dominant) and one short (recessive) factor, it will grow tall. For a recessive factor to be expressed, the plant needs two copies of it.

Mendel explained inheritance in terms of the two parents each contributing one factor to the offspring. If both are dominant, or one is dominant and one recessive, the offspring will show the dominant trait. If both are recessive, it will show the recessive trait. This was clear to him from the ratio of features in the second generation.

Inheritance of height in pea plants over two generations.

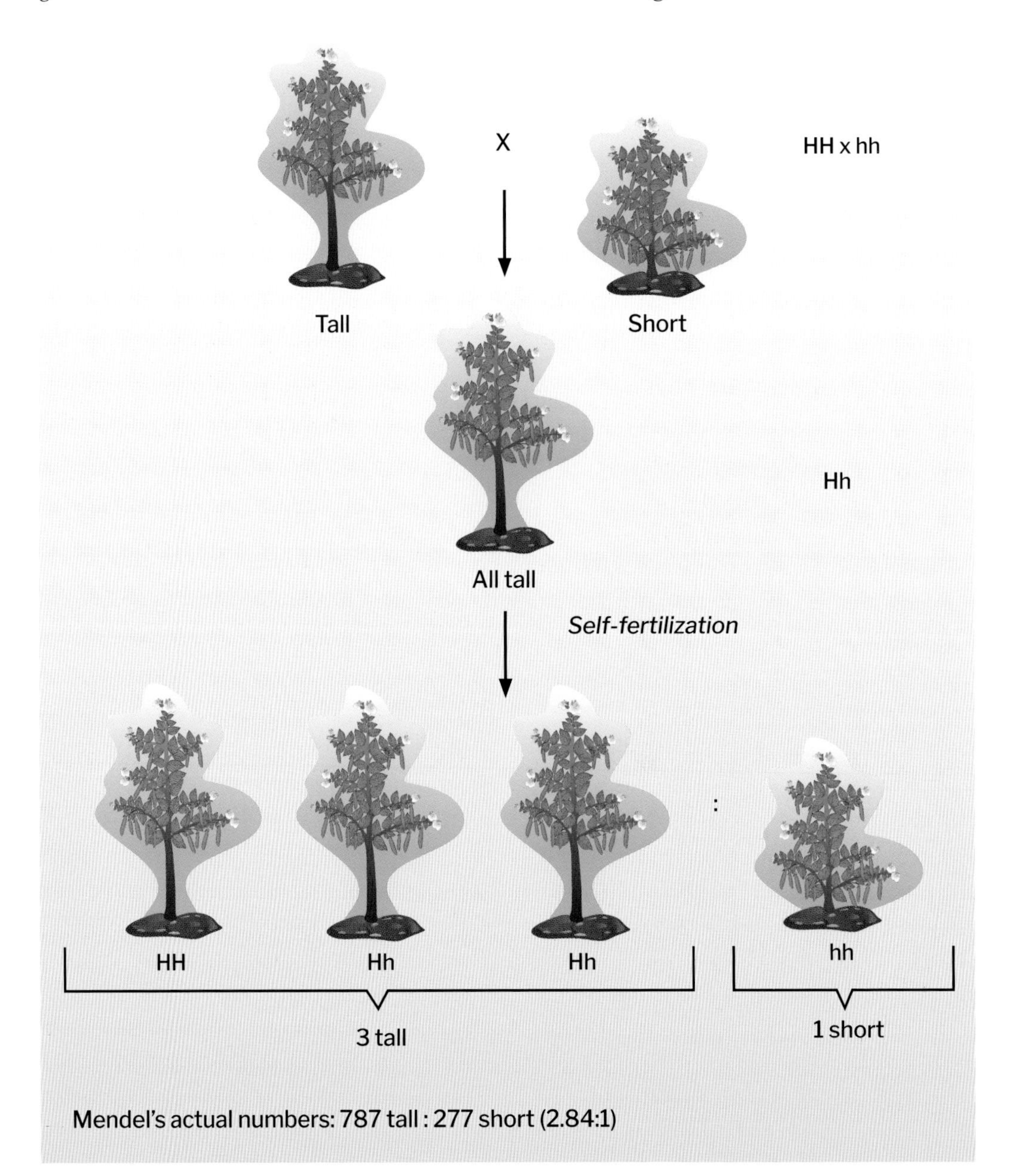

A mystery unravelled

We now know that the mechanism of inheritance is DNA, deoxyribonucleic acid. This long molecule, with the shape of a twisted ladder – called a double helix – carries genetic characteristics coded into its structure. Each molecule of DNA forms a chromosome. Each chromosome carries the coding for many different characteristics of an organism.

Mendel's factors of inheritance are now called alleles. Rephrasing Mendel's findings, a pea plant has a gene for height, which can have either of two alleles, tall (H) or short (h). Tall is dominant and short recessive. The gene for height is a particular chunk of one of the pea plant's chromosomes.

MIX AND MATCH

Mendel found that the features he was studying were inherited separately, so a plant might have the recessive trait for height but the dominant trait for flower colour, for example. This led him to conclude that the factors for different features are split separately when egg and sperm cells (called gametes) are created. Each new pea plant gets, separately, a factor for each feature from each parent. It's a matter of chance which of each parent plant's two factors goes into each gamete and so which a particular pea plant inherits.

Although Mendel had worked out the mathematics of inheritance, he had no idea of the mechanics of it. How exactly did an organism inherit features from its parents? That remained unknown. Mendel published his findings in 1866, with data from 30,000 pea plants, but for 35 years it lay unnoticed.

FROM MENDEL TO MORGAN

In 1869, just a few years after Mendel published his work, the Swiss chemist Friedrich Miescher discovered a substance in the nuclei of white blood cells which he named nuclein. It is now known as DNA. He realized it was important but had no idea what it did. No one made the connection with Mendel's work on heredity. Indeed, no one made any connections at all with Mendel's work until 1900 when three botanists independently rediscovered it and carried out experiments that supported his findings. It still didn't immediately appeal to evolutionary biologists as Charles Darwin had talked of evolution coming about through the blending of features of both parents, and clearly Mendel's model didn't allow any blending.

Soon after, the young American biologist Walter Sutton closely observed meiosis (the cell-division process that makes gametes) in grasshopper cells and realized that the chromosomes split up in exactly the way that Mendel's explanation required. He suggested in 1902 that chromosomes are how Mendelian inheritance works. A pair of chromosomes has two genes for each characteristic. When the chromosomes separate in meiosis, each chromosome has one gene for each characteristic. When an egg is fertilized, a corresponding chromosome contributed by the other parent brings the second gene. This is now known as the Sutton–Boveri theory of inheritance. (Theodor Boveri came up with much the same idea at the same time.)

Thomas Morgan demonstrated this conclusively through experiments with fruit flies. In 1911, he showed that particular genes are carried on particular chromosomes. Morgan's PhD student, Alfred Sturtevant, took this further, producing the first gene map, which shows where specific genes appear on chromosomes of the fruit fly.

By explaining the biochemical mechanism of inheritance, Mendel and those who came after him provided an explanation for evolution, but also opened the door for new developments previously undreamt of: genome sequencing, genetic engineering and genetic medicine.

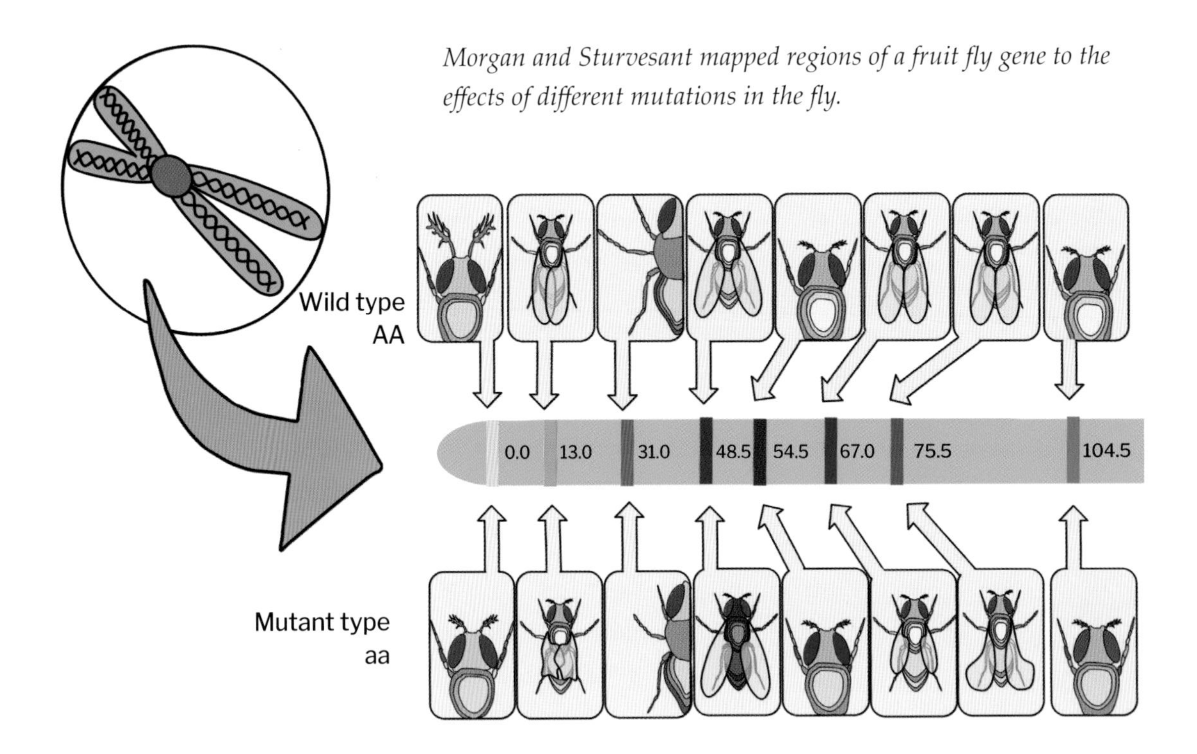

Morgan and Sturvesant mapped regions of a fruit fly gene to the effects of different mutations in the fly.

4

Cutting the uncuttables

Atomic structure

For centuries, everyone accepted that atoms are 'uncuttable' – it was part of their definition. But we now know that atoms are made of subatomic particles: one or more electrons in orbit around a nucleus that contains protons and – in all elements except hydrogen – neutrons. The protons and neutrons are themselves made up of quarks.

MAKING MODELS

The first subatomic particle discovered was the electron, found by J.J. Thomson in 1897. He determined that the 'cathode rays' physicists had been experimenting with were actually streams of tiny, negatively charged particles, much lighter than the lightest atom, and were components of atoms. It led to the first model of what the atom might be like in 1904. Thomson's 'plum pudding' model had a cloud of positive electrical charge studded with negatively charged electrons like a traditional English plum pudding: a blob of suet pudding studded with dried fruit.

A famous experiment in 1909 to test Thomson's model showed it was wrong. Hans Geiger and Ernest Marsden fired alpha particles (helium nuclei, with a positive charge) at thin gold foil. They expected most to pass straight through, with some being slightly deflected if their path took them close to an electron. The result was stunning: most did pass through, but some were considerably deflected and a few even bounced straight back. From this, Ernest Rutherford designed a new model of the atom in 1911. This had all the positive charge focused in a single point at the middle, which he soon called the nucleus, and the negative charge forming a much larger cloud around it. This model was refined by the Danish physicist Niels Bohr who constrained the electrons in specific orbits, rather like planets around a star (hence it's called the planetary model). The electrons could not stray between these orbits, though they could jump between them, which either required or released energy depending on the direction of the jump.

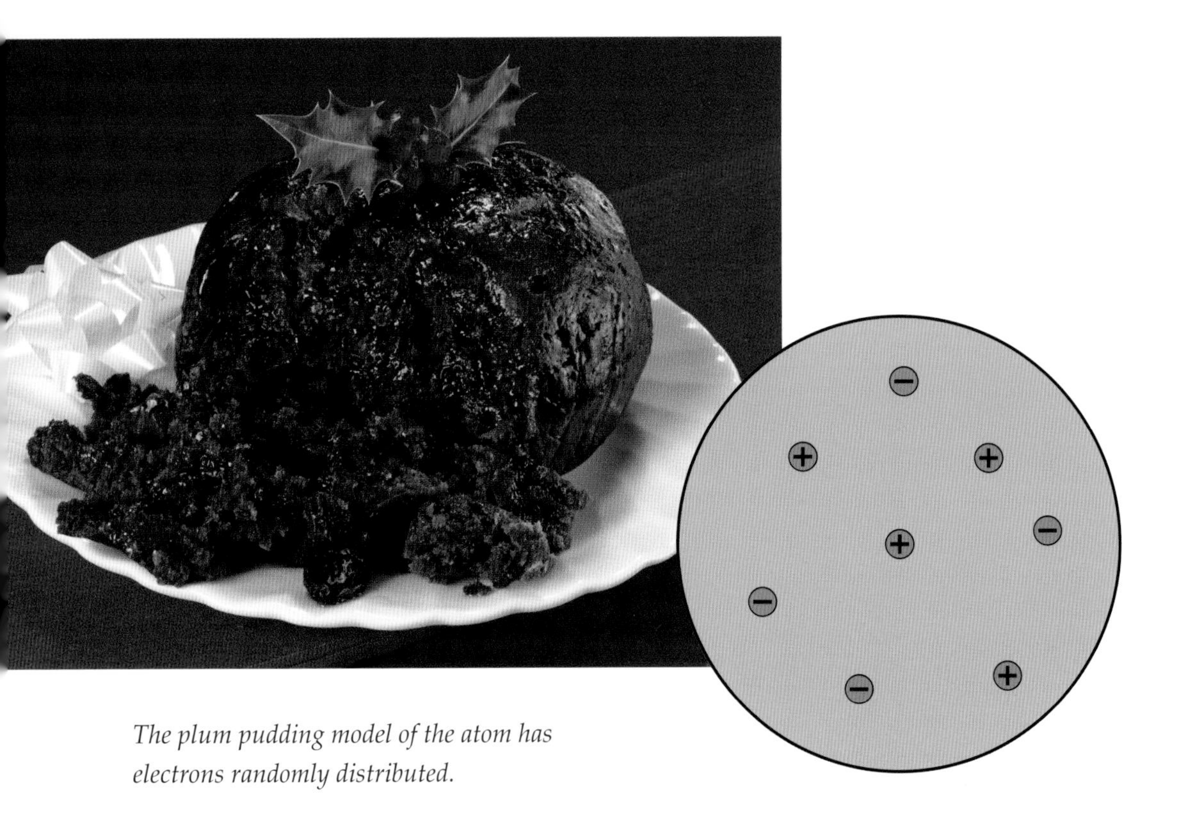

The plum pudding model of the atom has electrons randomly distributed.

BITS IN THE MIDDLE

The question of what formed the nucleus took a little longer to solve. The periodic table (see page 31) gave each element an atomic number which, it later emerged, was the same as the positive charge on the nucleus of the atom. The mass of an atom of hydrogen, though, must represent the mass of one proton. So how could an element with a mass of four (helium) have an atomic number of two? The assumption was that 'internal electrons' in the nucleus balanced the protons, giving the correct mass and the overall neutral charge of the atom. (This wouldn't work as the mass of an electron is negligible – a proton has 1,800 times its mass.) In fact, there are no internal electrons: the extra mass is made of uncharged particles, called neutrons, discovered by James Chadwick in 1932. A neutron has the same mass as a proton. A helium atom, then, has two protons, balancing the charge of the two electrons, and two neutrons which make up the rest of its atomic mass.

BITS IN THE BITS

The assumption that neutrons and protons are fundamental particles – so can't be broken down further – was challenged in the 1960s. As more and more tiny particles were discovered or theorized, the model needed revision. In 1964, Murray Gell-Mann proposed the existence of quarks (see page 128).

The current description of the atom has each proton and each neutron made of three quarks, though the electron remains complete in itself. The original 'uncuttables' have been thoroughly cut.

STICKING TOGETHER

Accepting that all matter is made of atoms raises another question: how do atoms stick together to make matter? Atoms form bonds with one another to make molecules. If the atoms are of the same element, the molecules are of that element, such as two hydrogen atoms making molecular hydrogen. If the atoms are of different elements, they form a compound, such as carbon dioxide (carbon and oxygen), or water (hydrogen and oxygen) or common salt (sodium and chlorine).

REDRAWING THE ATOM

Niels Bohr's atom had electrons each occupying a 'shell' around the nucleus. The electron shell is three-dimensional, like a sphere around the nucleus, rather than the two-dimensional, nearly circular orbit of a planet. Shells can hold different numbers of electrons, and an atom is most stable when its outer electron shell is full – it has as many electrons as it can hold. The innermost shell only ever holds a maximum of two electrons; the next shell holds eight. Within each shell are subshells, and within the subshells, individual electrons each occupy a three-dimensional space called an orbital.

ELECTRONS AND THE ELEMENTS

The atomic number of an element tells you how many protons an atom of the element has. It has the same number of electrons, so that the atom has neither a positive nor negative charge. If an atom gains or loses an electron, it is 'ionized' – it becomes an ion, with an electrical charge.

Elements in the periodic table are arranged in vertical groups. All elements in a group have a similar configuration of electrons in their outermost occupied electron shell. For example, the elements in group I have one electron in the outer shell; those in group VII have space for one more electron in their outer shell. The elements of group VIII, the noble gases, have complete outer shells.

Atoms are most stable when their outermost shells are filled, with all the orbitals occupied by an electron. They will form bonds with other atoms to achieve that type of stability, by gaining, losing or sharing electrons. By forming bonds with other atoms, they make molecules.

ELECTRONS GOING SPARE

There are two types of chemical bonds: ionic and covalent bonds. In an ionic bond, one atom 'gives' an electron to another atom. Both atoms become ions. An example is the salts formed by the group I elements (alkali metals) with the group VII elements (halogens). One atom of an alkali metal, such as sodium or potassium, donates an electron to an atom

Sodium and chlorine atoms become ions by trading an electron.

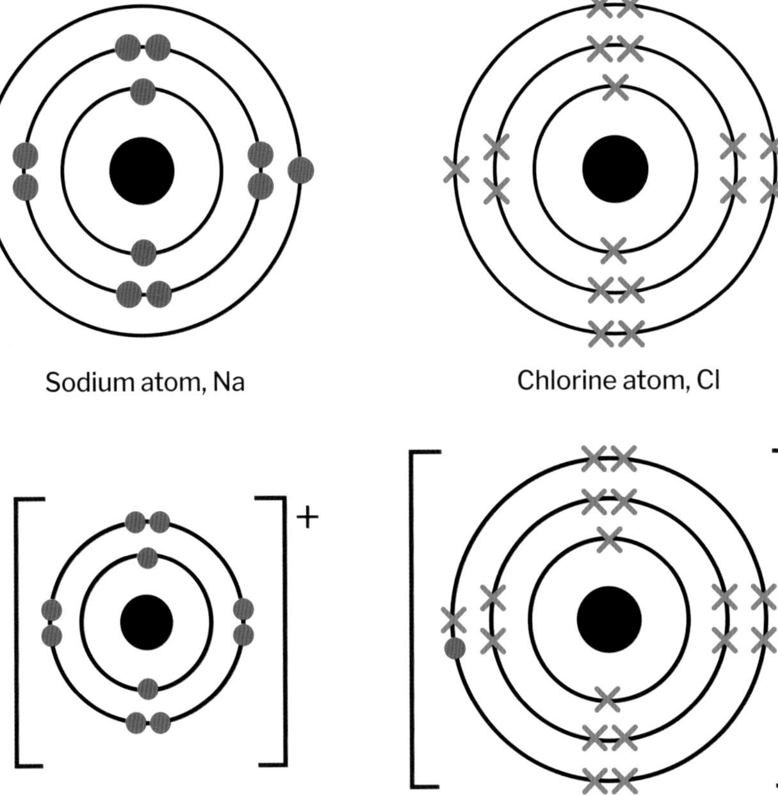

of a halogen (such as chlorine). The result is that the alkali metal gets rid of its spare electron, leaving it with all its occupied shells filled, and the halogen completes its outer shell, again achieving filled shells. The two elements are highly reactive as their atoms are unstable. Once they form a compound, they are stable. This is an ionic bond because each atom has become an ion: the alkali metal gives away its 'spare' electron to become a positively charged ion and the halogen takes it in, becoming a negatively charged ion.

Atoms can give away or accept more than one electron, becoming ions with positive or negative charges greater than one. For example, oxygen has eight electrons, two in the inner shell and six in the next shell, leaving two unfilled spaces in that shell. It accepts two electrons to become the oxide ion O^{2-}.

Ionic compounds form large lattice structures; in salts, we see these as crystals. The attraction between the positive and negative ions holds them together. It's difficult to overcome the

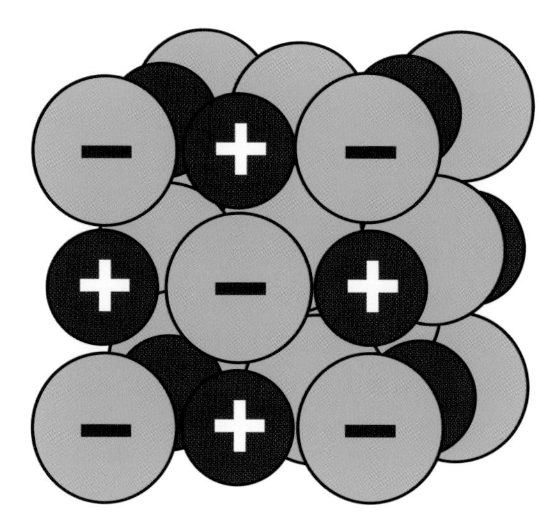

An ionic compound has a large lattice of positive and negative ions.

strong electrostatic forces that hold the ionic lattice together, so these compounds have high melting and boiling points and are always solids at room temperature. Sodium chloride melts at just above 800°C (1,472°F).

SHARE AND SHARE ALIKE

In a covalent bond, the atoms share a pair of electrons. Each atom contributes one electron to the shared pair. Instead of leaving their original atom, the electrons construct an orbital that allows them to be involved with both atoms. Bonds between non-metals, or between two metals, are often covalent. An example is hydrochloric acid, a compound of hydrogen and chlorine. Each contributes an electron, so hydrogen ends up with two shared electrons filling its only shell, and chlorine fills its outer shell by using the shared pair as electrons seven and eight. Water is formed from one oxygen atom sharing electrons with two hydrogen atoms. Each hydrogen atom has one pair of shared electrons, and the oxygen atom is sharing two pairs. As oxygen is two electrons short in its outer shell, this gives it a full set of eight electrons in that shell.

By looking at the position of an element in the periodic table and calculating the number of electrons in its outermost occupied shell, it's possible to work out how it's likely to combine with other elements to form simple compounds.

A water molecule involves two pairs of shared electrons, making covalent bonds.

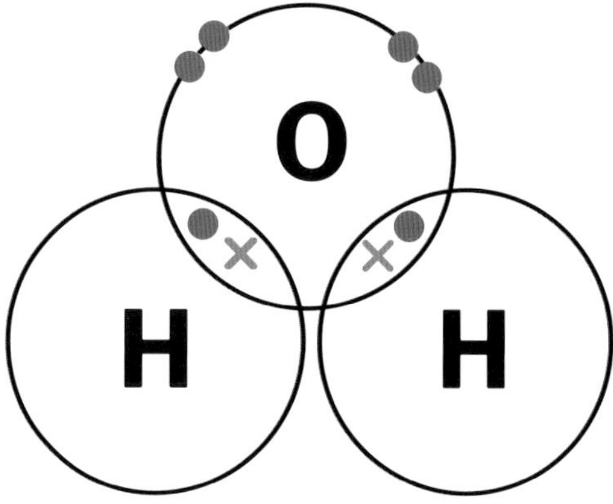

5

Painless surgery

Anaesthetics

Today, if you need a surgical procedure, you can reasonably expect to sleep through it, or at least have the appropriate part of your body numbed so that you don't feel it. That's one of the greatest luxuries of modern life. Until about 150 years ago, pain was the inevitable companion of most medical treatments.

PAIN'S NOT JUST A PAIN

Obviously, enduring terrible pain during an operation is an awful thing to contemplate, but pain itself is not the only problem with pain. A person in pain isn't going to lie quietly while a surgeon pokes around inside them, removes an arrow from their eye, or sets a broken leg. They're going to struggle and lash out and make life very difficult for the medical practitioner with the result that the procedure will be extremely hard or impossible to do properly. This meant that only straightforward, quick procedures were even possible.

Hua Tuo might have made an effective anaesthetic.

You could cut off a leg by holding the patient down, but removing an appendix or performing heart surgery would not be possible. Even the procedures that could be done might result in the patient's death from shock. The invention of anaesthesia made possible more complicated surgery that needed the patient to be still for an extended period.

POPPIES AND POTIONS

For hundreds of years people made drafts with varying efficacy from plants including poppies (the source of opium), or gave patients inordinate amounts of alcohol, or even hit them over the head – all of which could do more harm than good and were rarely very effective. There are reports of anaesthesia in ancient times, but insufficient details to know how effective they were. In China, Hua Tuo (*c.* AD 140–208) was said to have performed surgery using a general anaesthetic called mafeisan (boiled cannabis powder), but the recipe does not survive. The Indian physician Jīvaka, at one point personal physician to the Buddha, is also said to have used some kind of anaesthesia and performed surgery on the gut and even the brain, but again no recipe for his anaesthetic survives and its use seems to have died with him in the 5th century BC.

LAUGHING GAS – NOT SO FUNNY

The first really useful steps towards anaesthesia came in the late 18th century when chemists began exploring gases. Nitrous oxide, discovered in 1772 by Joseph Priestley, became the subject of study by Humphry Davy in 1798. Davy reported his findings on the gas he called 'laughing gas' in 1799, noting that inhaling it caused euphoria and also reduced feelings of pain – the pain from his wisdom teeth was dulled when he inhaled nitrous oxide. Although this looks like the perfect opportunity to launch anaesthetic gases into the world, Davy and his friends instead used it recreationally and gave demonstrations of its amusing effects at the Royal Institution. It was there, in 1813, that Michael Faraday joined him and explored the effects of ether, which again included pain relief and putting people to sleep. As one of his subjects took a full 24 hours to recover from the ether treatment, he didn't pursue it.

The entertainment value of nitrous oxide eventually brought it to the attention of an American dentist, Horace Wells, who did try to exploit it. After seeing a showman demonstrate the gas, he arranged to inhale nitrous oxide himself and had a colleague remove one of his teeth. Satisfied, he started using it with his patients, but when he gave a demonstration in Boston, his patient cried out. Although Wells continued to use it in his practice, he gave up trying to promote its use. His former student, and assistant at his demonstration, William Morton, took up the baton. Realizing nitrous oxide wasn't going to be good enough, he experimented with ether, using it first on animals and then patients before giving a demonstration in Massachusetts General Hospital. There, in 1846, he removed a tumour from the neck of an anaesthetized man who felt nothing. The age of pain-free surgery had begun.

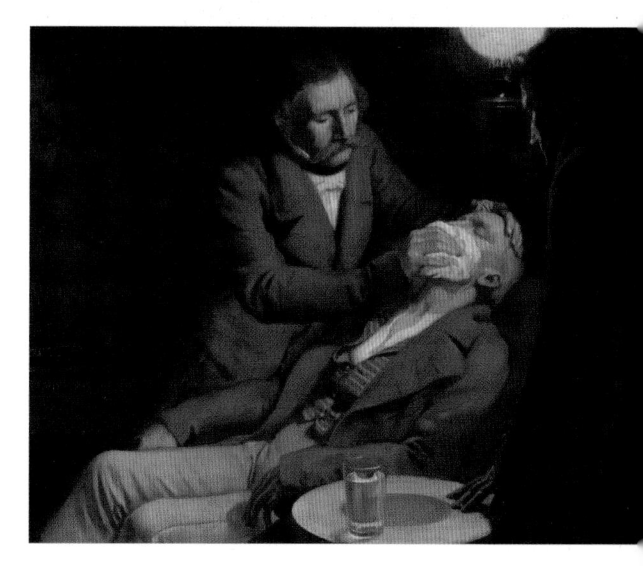

William Morton's first use of ether in dental surgery.

Six hundred years of pain

Di-ethyl ether, used by Morton, was first made in the 13th century, from sulphuric acid and ethanol. At the time, people noticed that it could relieve pain and even lead to unconsciousness – but it took six more centuries before anyone thought to put this to good medical use. It perhaps didn't help that people accepted that pain was the lot of humankind. In some religions, it was even considered a spiritual advantage, bringing the sufferer closer to God or indicating a special trial sent to favoured individuals. Maybe people simply weren't ready for the idea that pain could or should be mitigated.

MOVING ON

Ether was not ideal in an age of candles and gas lights as it's highly inflammable. Being heavy, it tends to lie close to the ground rather than dissipate and causes a fire risk. It has unpleasant side effects, including vomiting, and takes a long time to take effect. This led people to search for alternatives, the first of which was chloroform, introduced in 1847. Objections to anaesthesia were finally laid to rest when Queen Victoria used chloroform during the birth of Prince Leopold in 1853.

The search for alternatives continued, and continues to this day, with a wide variety of anaesthetics now differently administered. Improvements include the first use, in Germany in 1884, of a local anaesthetic (cocaine, in that case) that numbed just a part of the body while the patient remained conscious.

ANOTHER HURDLE

Anaesthesia produced patients who would lie still without complaining, who didn't remember what had been done to them and who were unlikely to die of shock. But that didn't immediately deliver a range of new surgical interventions. Opening a body in a dirty hospital, by a man whose clothes were spattered with the blood of his previous patients and whose instruments might have been wiped on his apron to clean them, carried considerable risk of infection. Most surgery patients still died. Another great innovation was needed to make surgery truly viable – antisepsis (see page 132).

A chloroform inhaler from France, c. *1860.*

6

Elementary

The chemical elements

As far as we know, everything in the universe is made up of just 118 chemical elements, each with their own unique type of atom. The idea that everything is made of a relatively small number of primary ingredients is very old, but it took a long time to get the current model of chemical elements.

FOUR OR MORE?

Several ancient cultures had theories of matter which involved four or more elements, but they were never substances we would now consider elements. In China, during the Spring and Autumn Period (770–476 BC), five elements were recognized. These were fire, wood, earth, water and metal. They were linked by 'generating' and 'overcoming' relationships in a complex web that was more philosophical than chemical. In Ancient Greece, in the mid-5th century BC, Empedocles described a system based on four 'roots' of matter, which was somewhat similar. The roots were earth, fire, water and air and they were subject to the two opposing forces of strife and love.

Alchemy was the forerunner of chemistry in China, the Arab world and Europe, the first form of systematic investigation of substances. The Islamic alchemists began with the four elements defined by the Ancient Greeks and built a highly complex system that mixed some genuine chemical knowledge with a lot of abstruse mystical and spiritual belief cloaked in a good deal of secrecy. They focused on the qualities and

The four elements.

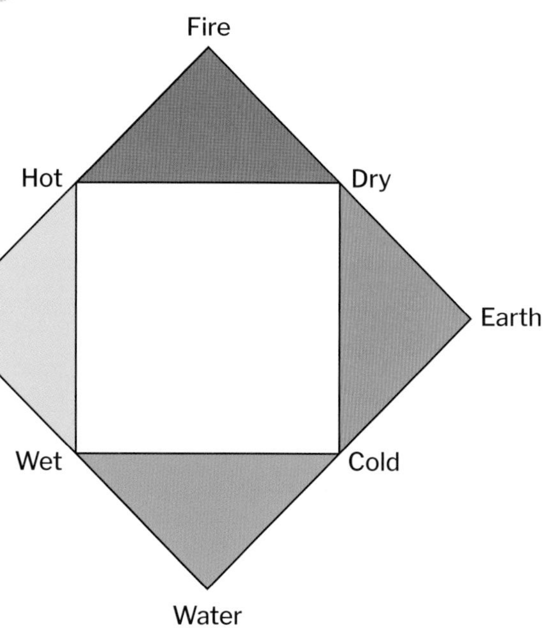

characteristics they thought the elements brought to matter. Later, alchemists promoted sulphur, mercury and salt as the elemental components of all metals. These were not the regular sulphur, mercury and salt we might find in any laboratory, but a kind of refined essence of the substances that bestowed properties on matter. All matter was thought to be made of the same essential substance, but with different properties which gave different metals different levels of purity. Rather confusingly, actual mercury was said to contain too much mercury! Gold was considered the purest substance, with silver coming second. One of the aims of alchemy was to turn base metals into gold, which alchemists believed could be achieved if the right mix of properties could be created by adjusting the levels of essential mercury and sulphur.

ELEMENTS BECOME CHEMICAL

For the alchemists, as for the Ancient Greeks, elements were not chemically distinct physical substances. The first person to suggest something comparable to the modern system of chemical elements was the French chemist Antoine Lavoisier (1743–94). He proposed that all matter is made from a restricted number of elemental materials that can combine in different proportions to make all other types of matter, a principle that still holds today. Lavoisier's initial list of 33 elements contains several substances still recognized as elements, a few now known to be compounds, and some that are not even considered substances now: light and 'caloric' (which was thought to make things hot). Lavoisier was careful to point out that he was listing substances that could not be further broken down in his day, but that they might be capable of further deconstruction in future laboratories, in which case they would no longer count as elements.

Lavoisier's definition of an element made no link to atoms. That connection was made by the English chemist John Dalton in 1801, who described each element as having its own style of atom, forming bonds on the basis of the features of its atoms.

MAKING MATTER

Chemical elements, such as gold or hydrogen, are made entirely of one kind

Antoine and Marie-Anne Lavoisier worked together, but Marie-Anne's contribution has been largely ignored by history.

of atom, but they don't all occur naturally in their atomic state. In some, the atoms form bonds with each other, so that the element exists in the form of molecules, such as molecular oxygen and molecular hydrogen. Each of these makes molecules of just two atoms (O_2 and H_2). Carbon forms giant molecules, with each carbon atom bonding with three or four others.

When atoms of different elements form bonds, the substance created is a compound. Examples include salt (sodium chloride) and sand (mostly silicon dioxide). Even though it might be very difficult to separate the elements in a compound, it is always possible. Any compound can be returned to its elements somehow, whether by burning, melting, dissolving or subjecting to some kind of chemical reaction. When Lavoisier noted that some of his supposed elements might one day be 'decomposed', he was recognizing the difficulty of breaking apart some compounds. A final type of substance is a mixture, in which elements and / or compounds are mixed together

Diamonds and coal

Some elements can arrange their atoms in different ways and so make substances which are physically different, even though all the atoms in them are identical. Carbon can exist as a dark grey solid, the graphite in pencil lead, or as diamonds. The different arrangements of the atoms give the two substances their different properties.

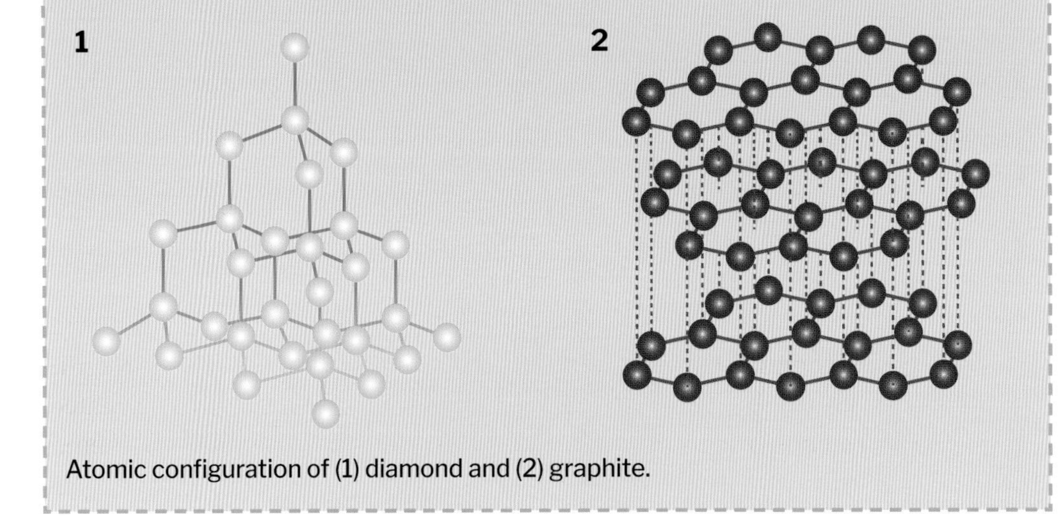

Atomic configuration of (1) diamond and (2) graphite.

but have no chemical bonds between them. An example is milk, which is a mix of water with various fats, sugars and proteins.

MAKING MORE ELEMENTS

More and more elements have been discovered since Lavoisier published his work in 1789. Most have been found naturally occurring, but some have been made in the laboratory. The first artificially synthesized element, technetium, was made in 1937. Radioactive elements decay, becoming other elements as the number of protons in the nucleus changes (see page 61). Emilio Segrè and Carlo Perrier found traces of technetium (atomic number 43) in a sample of molybdenum (42) that had been exposed to high-energy radiation. Some of the molybdenum had captured an extra proton in the atomic nucleus, so becoming technetium.

Some of the heaviest radioactive elements can only be synthesized in minute quantities for the shortest moments. Only five, or perhaps six, atoms of element 118, organesson, have ever been made and it's calculated to have a half-life of less than a millisecond. This means half of any sample of organesson would decay into another element in less than a millisecond, so there has not been time to study it.

7

Imposing chemical order

The periodic table

When Lavoisier proposed that all matter is made from a limited number of chemical elements, he produced a list in which order was not important. As more elements were discovered, and some of Lavoisier's list was discarded, it became apparent that some elements share similar features.

This suggested to chemists that perhaps the elements could be put into a meaningful sequence or groups. The task was successfully achieved by the Russian chemist Dmitri Mendeleev in 1869.

SIMILAR BUT DIFFERENT

Lavoisier had grouped 33 substances he considered to be elements by fundamental similarities: metals, non-metals, earths (or minerals) and a group of non-solids that included heat, light and three gases (hydrogen, oxygen and nitrogen). Some he suspected weren't elements, but he hadn't been able to break them down. Dalton's contribution to atomic theory, that in a given compound the elements always combine in the same ratios, was important in a different way. It enabled the calculation of the relative atomic masses of many elements, which hinted at a possible way of sequencing them.

A GAME OF PATIENCE

By the time Mendeleev tackled the

Dmitri Mendeleev.

problem of organizing the elements, 63 of them were known. He began by writing the name and atomic mass of each element on a piece of card, along with its properties. He then laid out all the cards in order of decreasing atomic mass. He found that some properties repeated at regular intervals. Chlorine and fluorine are similar, for example, as are sodium and potassium. He discovered that some features recurred periodically, and stated his law: 'If all the elements are arranged in the order of their atomic weights, a periodic repetition of properties is obtained.'

MIND THE GAP

Mendeleev drew up his version of the periodic table in 1869. Although we are used to seeing the table with groups arranged vertically and periods horizontally, his original table had the opposite orientation. The seven main groups he identified formed the longest rows. Above and below this he put the shorter rows of elements we now think of as transition elements. Confident that he had identified genuine periodicity, he left gaps for elements that he felt should exist, noting their predicted atomic mass and some of their properties. He also, occasionally, reversed the order of pairs of elements to keep the trends of properties going even at the cost of disrupting the sequence of atomic masses.

ОПЫТЪ СИСТЕМЫ ЭЛЕМЕНТОВЪ.

ОСНОВАННОЙ НА ИХЪ АТОМНОМЪ ВѢСѢ И ХИМИЧЕСКОМЪ СХОДСТВѢ.

			Ti—50	Zr=90	?—180.
			V=51	Nb—94	Ta—182.
			Cr—52	Mo=96	W—186.
			Mn—55	Rh—104,4	Pt=197,4.
			Fe=56	Rn—104,4	Ir=198.
			Ni—Co=59	Pl=106,6	O-=199.
H=1			Cu—63,4	Ag—108	Hg—200.
	Be=9,4	Mg—24	Zn=65,2	Cd=112	
	B=11	Al=27,4	?—68	Ur=116	Au—197?
	C=12	Si—28	?=70	Sn=118	
	N=14	P—31	As—75	Sb=122	Bi=210?
	O=16	S=32	Se—79,4	Te=128?	
	F=19	Cl=35,5	Br=80	I—127	
Li=7	Na=23	K=39	Rb=85,4	Cs=133	Tl—204.
		Ca—40	Sr=87,6	Ba—137	Pb=207.
		?—45	Ce—92		
		?Er—56	La=94		
		?Yt—60	Di—95		
		?In—75,6	Th—118?		

Д. Менделѣевъ

Mendeleev's first periodic table.

FALLING INTO PLACE

As new elements were discovered that matched Mendeleev's predictions and slotted into his periodic table, his ideas were confirmed and became more widely accepted. In some cases, where he had put elements apparently out of sequence for their atomic weight, their atomic weight was later recalculated and fitted the position he had given it. By 1890, the table was accepted as valid.

Mendeleev's confidence that led him to predict missing elements has contributed to him being widely acknowledged as the originator of the periodic table. In fact, there was a dispute over priority with the German chemist Lothar Meyer. Meyer had created a table in 1864 and perfected it in 1868. It showed 28 elements ordered by atomic mass and split into groups that reflected their valency. Valency is an atom's combining

Pair swaps

It might seem as though swapping pairs of elements completely violates the principle of the periodic table, but it turned out to be entirely legitimate. Iodine and tellurium are a pair that Mendeleev switched. Iodine has an atomic mass of 127, but tellurium exists in several isotopes. Isotopes of an element all have the same number of protons and electrons, but they have different numbers of neutrons. This means isotopes have different atomic masses. The isotopes of tellurium that are most abundant have atomic masses of 128 and 130, making the atoms heavier, on average, than iodine – but they all have fewer protons and electrons than an atom of iodine and should indeed come before iodine in the periodic table.

power, reflecting the ratio in which it combines with other elements. So oxygen has a valency of two (H_2O, water, has two hydrogen atoms combining with one oxygen atom) and carbon has a valency of four (CH_4, methane, has four hydrogen atoms combining with one carbon atom). We now know that valency reflects the occupancy of the outermost electron shell, but valency was studied before the existence of electrons was recognized. Meyer published his table in 1870, changing the vertical/horizontal arrangement, so that the groups (indicated by valency) were then horizontal. Mendeleev made the opposite change, having started with the groups horizontally and rearranging to make them vertical.

Meyer's work is certainly comparable

Lothar Meyer's periodic table, drawn up in 1864.

Valence	4	3	2	1	1	2
Element	–	–	–	–	Li = 7.03	Be = 9.3
	C = 12.0	N = 14.04	O = 16.00	F = 19.00	Na = 23.05	Mg = 24.00
	Si = 28.5	P = 31.0	S = 32.07	Cl = 35.46	K = 39.13	Ca = 40.00
	–	As = 75.0	Se = 78.8	Br = 79.97	Rb = 85.4	Sr = 87.6
	Sn = 117.6	Sb = 120.6	Te = 128.3	I = 126.8	Cs = 133.0	Ba = 137.1
	Pb = 207.0	Bi = 208.0	–	–	Tl = 204	–

with that of Mendeleev, but Meyer was less vocal in promoting his scheme, included far fewer elements, and also did not publish it until 1870 – after Mendeleev had revealed his own. Meyer also held back from predicting missing elements and instructed his students using his table not to use it predictively.

ALL NUMBERED

Mendeleev numbered the elements in his sequence, and for a long time the atomic number of an element seemed to relate only to its position in the list. As it turns out, the atomic number is a good description of the structure of an element's atoms. The number corresponds to the number of protons or electrons (which are the same unless it's ionized). The group (vertical column) in which the element is found indicates the number of electrons in the outer shell of the atom. The period (horizontal row) in which the element is found indicates which is the outermost occupied electron shell of the atom. For example, potassium is element 19; it has 19 electrons and 19 protons. Its electron configuration is 2.8.8.1, meaning it has two electrons in the innermost shell, eight in each of the next two shells, and only one in the outermost occupied shell. From this we can tell that potassium is in group I (it has one electron in the last shell) and in period 4 (the final electron is in the fourth shell).

MORE ELEMENTS

The noble gases (group VIII) were unknown to Mendeleev, but the addition of an entire group to the end of the rows was not very disruptive. The recognition that the transition metals – in ten shorter columns, from scandium to zinc – had to be accommodated between groups II and III required redrawing the table with its now-familiar profile, while Mendeleev had put them originally either side of the block of groups I–VII. The lanthanides and actinides were fitted into this expansion later.

The periodic table has grown from Mendeleev's original 63 elements to the 118 elements now known. The heavy 'transuranic' elements (heavier than uranium) are all radioactive and many have very short half-lifes (see page 62). They aren't found in nature as they disappear by radioactive decay almost as soon as they are created. They have been created in the laboratory as fleeting presences in a decay chain, with half-lifes of tiny fractions of a second. The heaviest element so far known is oganesson, with a half-life of less than a millisecond.

ARE THERE MORE?

Chemists around the world have tried to find elements beyond 119, which would mean adding a new period to the table. So far, they have not succeeded. As an atom gets larger, its electrons have to move more quickly. Some calculations suggest that will be the limiting factor on heavier elements. An electron can't go faster than the speed of light. Some calculations put that limit at element 137, others at 170.

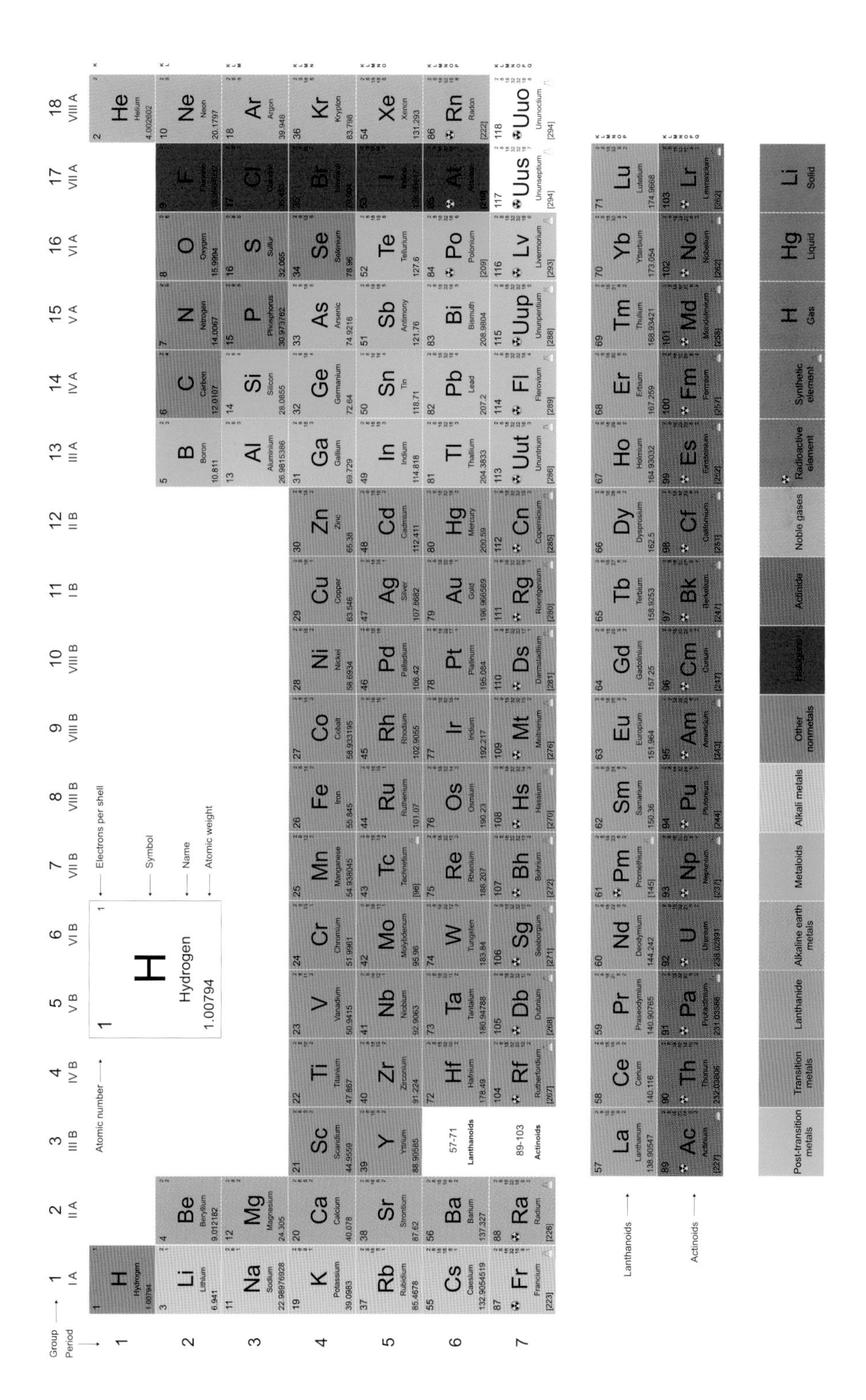

The modern periodic table.

8

All the way down

Gravity

It's fairly obvious that things fall downwards, but that's not what gravity is really about. The story of Isaac Newton thinking of his theory of gravity when an apple fell from a tree while he sat in the orchard of his family's Lincolnshire farm in 1665 doesn't mean that he noticed gravity makes things fall down.

It means that he wondered – and then worked out – why things fall downwards rather than sideways or in any other direction. What actually causes the trajectory of falling objects? Newton's answer explains how the universe is held together.

DOWN, NOT ALONG

Newton's explanation of gravity doesn't include up and down at all. Instead, gravity is explained as a force that operates between two objects that have mass. Gravity draws an object towards the centre of mass of another object. The centre of mass of the planet Earth is the middle of the planet, deep within.

Wherever you are on Earth, anything you drop is attracted to the centre of the planet, so moves in the direction we call 'down'. Clearly, an object dropped at the North Pole, one dropped at the South Pole, and one dropped at the equator all fall down, but viewed from outside Earth they're moving in different directions.

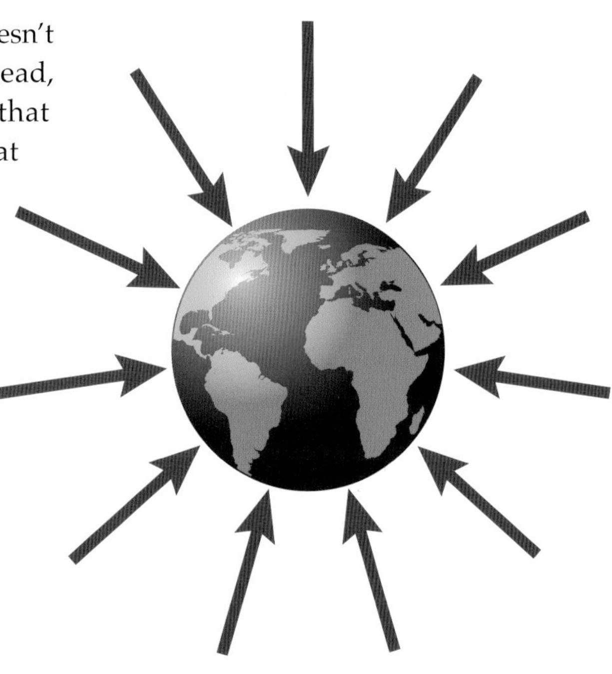

Earth's gravity pulls objects towards the centre of the planet.

$$\text{Force} = \frac{\text{Gravitational constant x mass small object x mass large object}}{\text{Distance between objects x distance between objects}}$$

Newton proceeded to work out the extent of the force of gravity, coming up with the formula:

$$F = \frac{GmM}{r^2}$$

Where:

F is the force operating between two objects (we measure this in Newtons; a Newton is 1 kg m/s2, which is the force which gives a mass of 1 kilogram an acceleration of 1 metre per second, per second)

G is the Universal Gravitational Constant (which is not actually 'universal' as it has a different value on different planets)

m is the mass of the smaller object

M is the mass of the larger object

r is the distance between the objects.

The force is felt by both objects. That means that you exert a gravitational pull on Earth, just as Earth exerts a gravitational pull on you. Obviously, it has less impact on Earth than it has on you.

FALLING, BUT HOW FAST?

Newton also defined force as mass × acceleration (F = ma). Substituting this in his gravity equation gives us a way of working out the acceleration of an object falling under gravity. The most

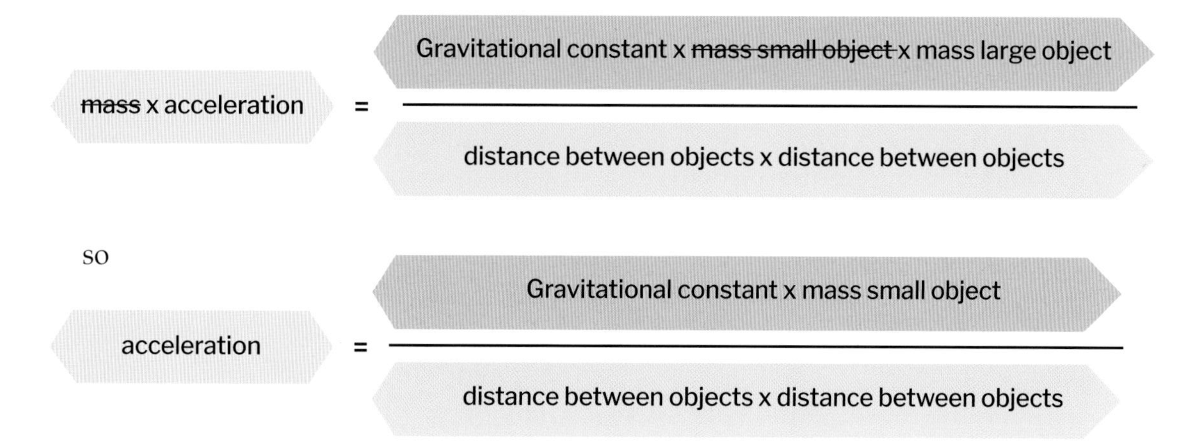

Isaac Newton.

interesting result of this – which surprises many people even today – is that the mass of a falling object doesn't affect the speed at which it falls. The mass of the small object, appearing on both sides of the equation cancels out – it has no effect. Galileo had already spent some time trying to demonstrate that objects of different mass fall at the same rate the previous century.

Everyday observation sometimes suggests that light objects fall more slowly than heavy objects, but this is the effect of air resistance, or drag, rather than gravity. Items of the same shape fall at the same speed, regardless of their mass. A flat piece of paper falls more slowly than a stone, but it also falls more slowly than the same piece of paper screwed into a ball. That mass really has no effect on the rate of falling was demonstrated on the Moon by astronaut David Scott in 1971, dropping a hammer and a feather. With no air to slow the descent of the feather, they both hit the surface of the Moon at the same time.

EXPLAINING SPACE

Newton used his theory of gravity to explain how the planets orbit the Sun and moons orbit the planets. When an object moves, it has velocity (speed) and direction. If either the velocity or the direction changes, the object is accelerating. It can only accelerate if a force acts on it to change either its velocity or direction. If there is no force acting on an object it will either not move (if it's stationary) or it will continue moving in the same direction and at the same speed. A planet going around a star is not moving in a straight line; it's constantly

Mass and weight

We're used to using 'mass' and 'weight' interchangeably and on Earth they have the same value. But in physics, mass is a measure of how much matter makes up an object while weight is a measure of the force of gravity acting on that mass. As long as you stay on Earth, they are effectively the same. If you lose 5 kg (11 lb) of body mass you will weigh 5 kg less. But if you went to the Moon, your weight would be less than it is now (without losing 5 kg) even though your mass would be the same. That's because the gravitational force of the Moon, which has lower mass than Earth, is lower than the gravitational force of Earth.

changing direction as it follows a nearly circular path. A force must be operating to change its direction, and that force is gravity. Gravity works to pull the planet towards the star, but an orbiting planet travels at the right speed to stay in orbit. The force of gravity isn't enough to pull it into the star, but is enough to prevent it escaping into space in a straight line.

Objects in a small orbit (close to the central object) have to move more quickly to maintain their orbit than those in a large orbit. Earth travels at nearly 30 km (18 miles) per second as it orbits the Sun, but Saturn, much further away, travels at only 10 km (6 miles) per second. Today, we can use Newton's laws and calculations to work out how to keep artificial satellites in orbit around Earth so that they don't either get pulled towards the surface and crash or escape into space.

A two-dimensional visualization of the curvature of spacetime.

GRAVITY UPDATED

Newton's theory of gravity served science perfectly well for more than 200 years. But in the early 20th century Albert

Arthur Eddington's photo of the 1919 eclipse.

Einstein took a new look at gravity and re-described it. Einstein saw gravity as curvature in the spacetime continuum. The usual analogy here is to think of a blanket pulled taut and a heavy ball placed on the blanket. The ball makes a dip in the surface of the blanket. If you put a smaller ball on the blanket, it will roll towards the dip. This shows the effect of gravity, but gravity works in three dimensions, not the two dimensions of a blanket.

It might sound as though there is little difference in practical terms between the two models – a force acting between objects with mass and curvature of spacetime. But there is an important difference in that curvature affects even things without mass. Einstein predicted that gravity would also affect the path of light. Einstein was proved correct by measurements made during a total lunar eclipse in 1919. The result, that light is bent by gravity, confirmed his theory of general relativity (see page 186). During an eclipse, stars that seem to be near the Sun can be seen because the sky is dark. Usually during daytime, we can't see any stars as the Sun is too bright. By comparing the apparent position of specific stars seen from two different places on Earth during the eclipse, Arthur Eddington and Frank Dyson confirmed that light was indeed being bent by the Sun to the extent that Einstein predicted.

HOLDING THE UNIVERSE TOGETHER

The force of gravity has built the stars and planets and even galaxies as well as now holding them in place. Gravity pulls objects with mass towards each other, and this has allowed planets to accrete from tiny particles of dust and ice coming together (see page 67). It has allowed stars to form by drawing gas molecules together, and then pulling them so close to each other as the mass of the clump increases that eventually nuclear fusion starts (see page 94). Gravity holds the atmosphere around our planet, and works against the power of dark energy to tear the universe apart (see page 100).

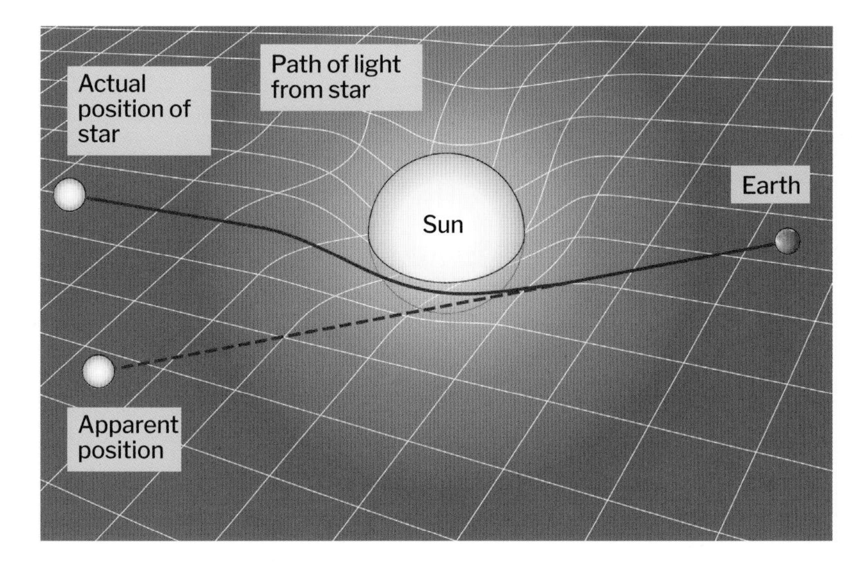

Even light, which has no mass, follows the curves of spacetime.

Gravity is one of the four fundamental interactions recognized by physicists. It is the weakest, but also the only one that has significant impact on macroscopic objects (those visible without a microscope). It works even at great distances, though its impact decreases following an inverse square rule: when the distance doubles, the effect of gravity falls to a quarter. It's not yet known whether gravity is quantized – that is, can be measured in discrete quantities – and mediated by hypothetical particles known as 'gravitons'.

If gravitational waves exist, mediated by particles of gravity (gravitons), the waves emitted by two merging neutron stars would look like this.

9

Starting with a Bang

The Big Bang

It's now widely accepted that the universe began as an infinitesimally small point that exploded into matter and spacetime around 13.77 billion years ago. This is called Big Bang theory; but the theory is only around 100 years old.

FUZZY START

The work that led eventually to Big Bang theory began in 1758 with a French astronomer, Charles Messier. He made a catalogue of 'nebulae' – fuzzy patches of light in the night sky that looked rather cloud-like (*nebula* is Latin for 'cloud'). His intention was to help his own quest for comets, as a catalogue of known fuzzy patches would make it easier to spot any new ones that might be comets. He added to the catalogue throughout his life, and his list numbered 103 by the time he died in 1817. There are now 110 'Messier objects'. They include the remnants of supernova events, star clusters and other galaxies.

HOW MANY GALAXIES?

Fast-forward to the 20th century, and some of Messier's objects attracted the attention of some of the century's greatest astronomers. In 1920, Harlow Shapley and Heber Curtis held a public debate on the size of the universe. At the time, most people assumed the Milky Way – the galaxy our solar system is in – was the whole of the universe. Shapley supported this view, arguing also that the Milky Way is very large, and the solar system is not near its middle. Curtis, on the other hand, argued for a much smaller Milky Way with the solar system central to it, but a host of other 'island universes' (galaxies) outside it. As it happens, Shapley was right that the Milky Way is very large and the solar system is not central, but Curtis was right about the existence of other galaxies. The debate didn't reach consensus, but the question was answered by Edwin Hubble three years later.

With better telescopes, the fuzzy patch in the constellation Andromeda resolved into a spiral shape. In 1923, Hubble could see that Andromeda contained stars. He used a method developed by Henrietta Swan Leavitt and published in 1912 to work out how far away they were – and the answer was

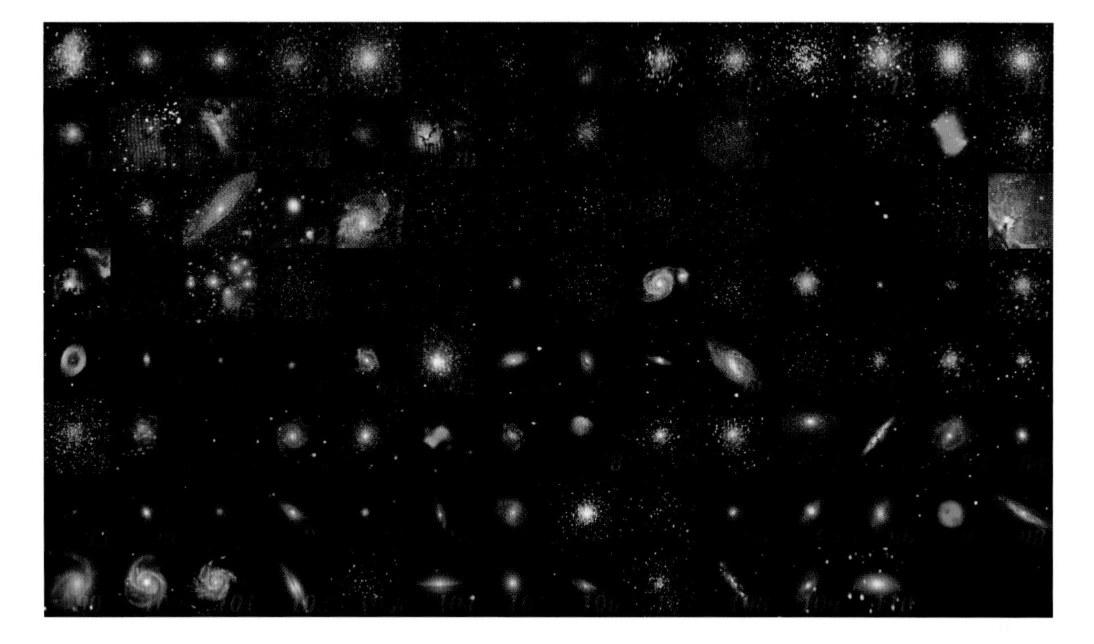

Photos of many of the Messier objects.

astonishing. Hubble calculated that the stars he could see in Andromeda were 900,000 light years away. (In fact, they are even further, at around 2.5 million light years distant.) This was far larger than even the largest estimates for the size of the Milky Way, so they must, after all, lie in an 'island universe'.

NEARER AND FURTHER

If the Andromeda galaxy were simply standing still relative to Earth, we might not have got any further. But in 1914, American astronomer Vesto Slipher presented his finding that most nebulae are moving away from Earth. He had calculated the redshift or blueshift for each object. This is the amount by which light from a distant object is shifted towards the blue or red end of the spectrum, a result of the object moving away from (red) or towards (blue) the observer. It's similar to the Doppler effect that makes the sound of a siren change as it approaches you, passes and then recedes into the distance. As space stretches between two objects, light passing between the objects has its wavelength stretched, so becomes more red.

FROM POINT TO POINTS

In 1927, the Belgian priest and astronomer Georges Lemaître put together Hubble's calculated distances to stars beyond our galaxy and Slipher's work on red- and blueshifted objects. He discovered that the most distant objects are moving away from us at the highest speeds. Lemaître realized this meant that the universe is expanding. If it has always been growing, he reasoned, it must have previously

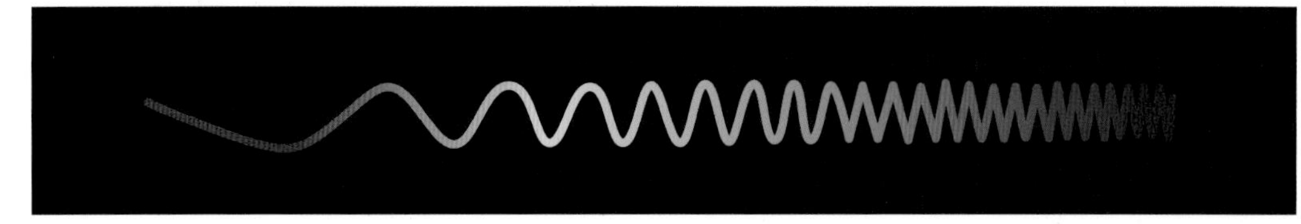

Light redshifted (stretched) by expansion of space.

been much smaller, and originally infinitesimally small. He called this first speck a 'primaeval atom' and declared it to be the origin of everything in the universe now. The basic matter and energy of the universe came into being in an instant, and it has evolved and spread out ever since. This theory is now called 'Big Bang' theory – a name first used by British astronomer Fred Hoyle in 1949 to mock it.

13.8 BILLION YEARS OF HISTORY

A lot had to happen to get from an infinitely small, infinitely dense point to the current state of the universe. Current thinking is that the universe began with a period of 'cosmic inflation' immediately after the Big Bang, which lasted only a tiny fraction of a second. By the end of it, the universe had grown from much smaller than an atom to about the size of a grapefruit, multiplying to 10^{26} times its original size (that's 1 followed by 26 zeroes). By the end of the first second, electrons, bosons, neutrinos, neutrons and protons (hydrogen nuclei) existed. After 20 minutes, neutrons and protons had combined to make the nuclei of helium and lithium. It would be nearly 300,000 years before these nuclei could capture electrons and become atoms, though. At that point, the universe became transparent as photons (little parcels of light energy) could disentangle themselves from a soupy mix of electrons and other particles and stream freely through the universe. This first flash of energy can still be detected as the cosmic microwave background radiation (CMBR). But there was nothing to see, even with photons free to move. It took at least 100 million years for the first stars to appear and start pouring out light.

Georges Lemaître.

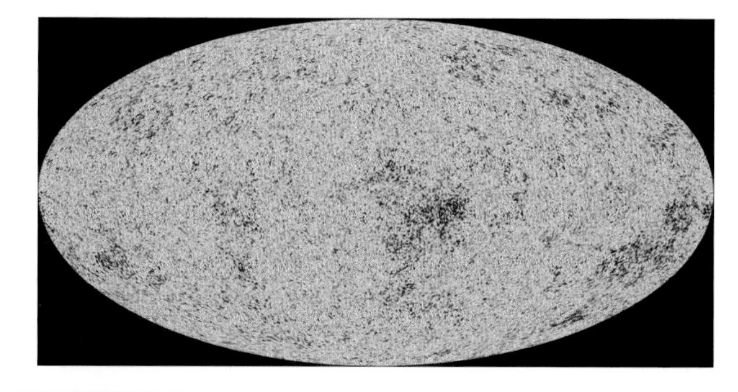

A map of the cosmic microwave background radiation reveals the uneven density of the universe.

The echo of a bang

The faint echo of the Big Bang can be detected as microwave radiation from all over the universe. It emerged as orange light when photons began to move freely and has been losing energy ever since. As it loses energy, the wavelength gets longer, so that it's now in the microwave area of the electromagnetic spectrum. Its existence was first suggested in 1948 and it was detected in 1964. It's considered important evidence of the Big Bang.

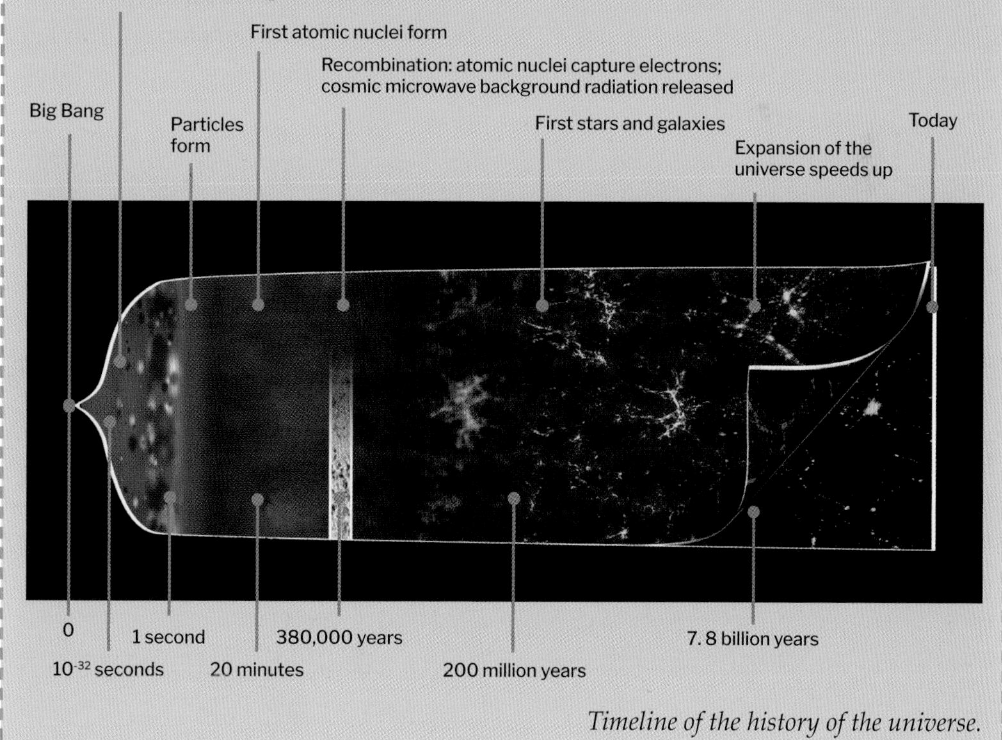

Timeline of the history of the universe.

10
Units of life
Cell theory

Every living thing is made of cells – tiny, self-contained pockets that act as little biochemical processors. The first living things on Earth were very simple organisms with a single cell of a very basic type **(see page 90).**

Every living thing on Earth has evolved from those cells and is still made of two fundamental types of cell that give us everything from bacteria to redwood trees and whales.

UNSEEN CELLS

No one knew of the existence of cells until after the invention of the microscope on the cusp of the 17th century. It would be easy to say that it's because cells are so small, but not all are microscopic. All eggs are a single cell, even the huge eggs of ostrich and crocodiles. Most cells, though, are very tiny.

The first person to see, describe and name cells was the English microscopist Robert Hooke. His astonishing book of items drawn through the microscope, *Micrographia*, revealed the microscopic realm to the public for the first time. Published in 1665, it included a drawing of the 'pores' or holes he saw in a piece of cork, which he described as looking rather like the spaces in honeycomb, but less regularly shaped.

With a nifty bit of maths, Hooke calculated that every cubic inch (16 cubic cm) of cork had nearly 1.3 billion cells. It must have been a mind-blowing moment when he realized the near-infinite complexity, at a microscopic level, of all the straightforward-looking things around him. The tiny compartments reminded him of the cells monks lived in in a monastery, and so he named them cells.

> *'[These] were indeed the first microscopical pores I ever saw, and perhaps, that were ever seen, for I had not met with any Writer or Person, that had made any mention of them before this.'*
>
> Robert Hooke

LIVING CELLS

Hooke had been examining cells that were long dead and empty – really, he saw only the walls they left behind. Cork

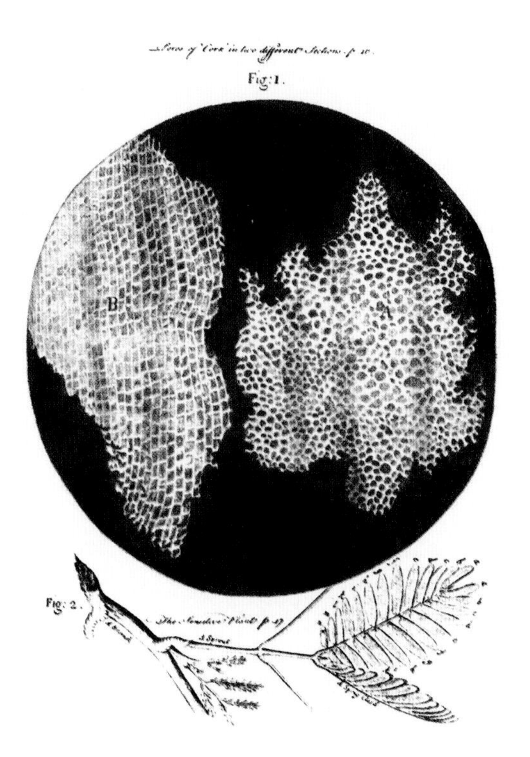

Hooke's drawing of cork cells.

is the bark of the cork tree, and the piece Hooke looked at had been stripped from the tree and processed into a plug for a wine bottle; it was very far from a living organism.

As microscopes improved over time, people began to see the true complexity of living cells. A cell has a barrier to separate it from the outside world, through which some substances can pass (the cell has some control over this). Inside is a gloopy fluid called cytoplasm and a collection of structures with different functions. The precise contents of a cell depend on its type, but all cells have: an outside; cytoplasm; genetic material (DNA, which contains its genetic information); and ribosomes, which can make proteins.

Although Hooke suspected that other living things were also made of cells – commenting in wonder 'how infinitely smaller then must be the Vessels of a Mite' – the theory that cells are the building blocks of all plants and animals was put forward in 1839 by Theodor Schwann. Schwann had worked extensively on cells in animals and had compared notes with Matthias Schleiden, who had worked on cells in plants. Each man had come to the conclusion that all living things in his domain (animals or plants) were made of cells. Schwann went on to propose the first main points of cell theory: all plants and animals are made up of cells, and cells are the building blocks of life. Rather ungallantly, he didn't credit Schleiden. In 1858, Rudolf Virchow added the final point of classical cell theory: that all cells come from other cells. That is, cells reproduce to create more cells and they are not created from inanimate matter.

BEYOND LIFE?

Scientists don't all agree on whether viruses count as alive. They don't have cells, and might be called packets of self-copying chemicals rather than living things. They can only reproduce inside a host cell – so in another organism, such as a plant, animal or bacterium. Something like viruses was probably a stepping-stone on the way to living cells, but it's debatable whether they can claim their 'living thing' badge.

Us and them: eukaryotic and prokaryotic cells

There are two very different categories of cells: eukaryotic and prokaryotic cells. The first type to evolve, and the simplest, is prokaryotic cells. Prokaryotic cells are always single-celled organisms: bacteria and archaea (a very old type of microorganism, still around and found living even in extreme environments). Everything else living is made of one or more eukaryotic cells.

Prokaryotic cells have a cell membrane around the outside, and can also have a rigid cell wall and a sticky carbohydrate layer called a capsule. Inside, they don't have separate structures with membranes around them. Instead, their bits and pieces float around in the cytoplasm. These consist of a loop of DNA, which holds all the genetic information defining the organism and how its cell functions, and the ribosomes, which are miniature factories that make proteins for the cell's functions.

Eukaryotic cells are souped-up cells. Inside, their structures have membranes around them to separate them from the cytoplasm. These

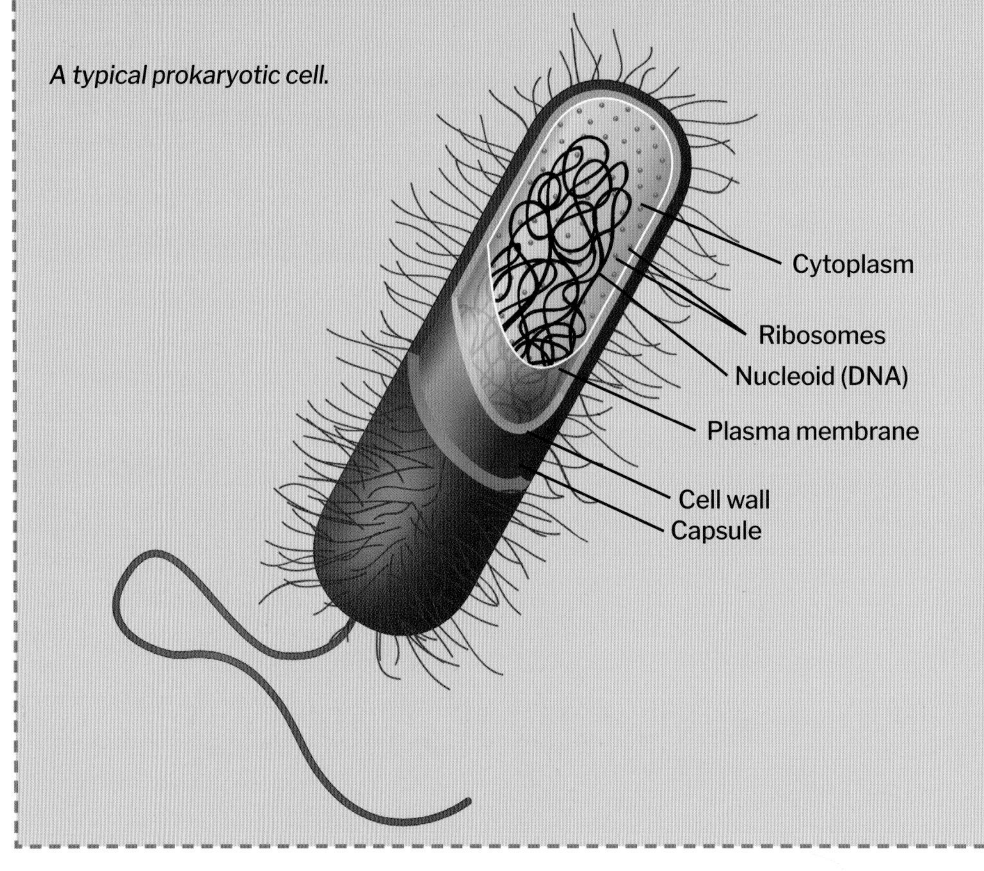

A typical prokaryotic cell.

structures are organelles – tiny organs that perform different functions within the cell. The type of organelle varies with the sort of cell. Green plants have organelles called chloroplasts which carry out photosynthesis, for instance, but animals don't have those. (There is more about organelles on page **146**.) The organelles carry out tasks such as storing and releasing chemical energy (in the mitochondria), making proteins (endoplasmic reticulum), packaging proteins and sending them to the right place in the cell (Golgi apparatus), and controlling all the activities of the cell (nucleus). The DNA is stored in the nucleus.

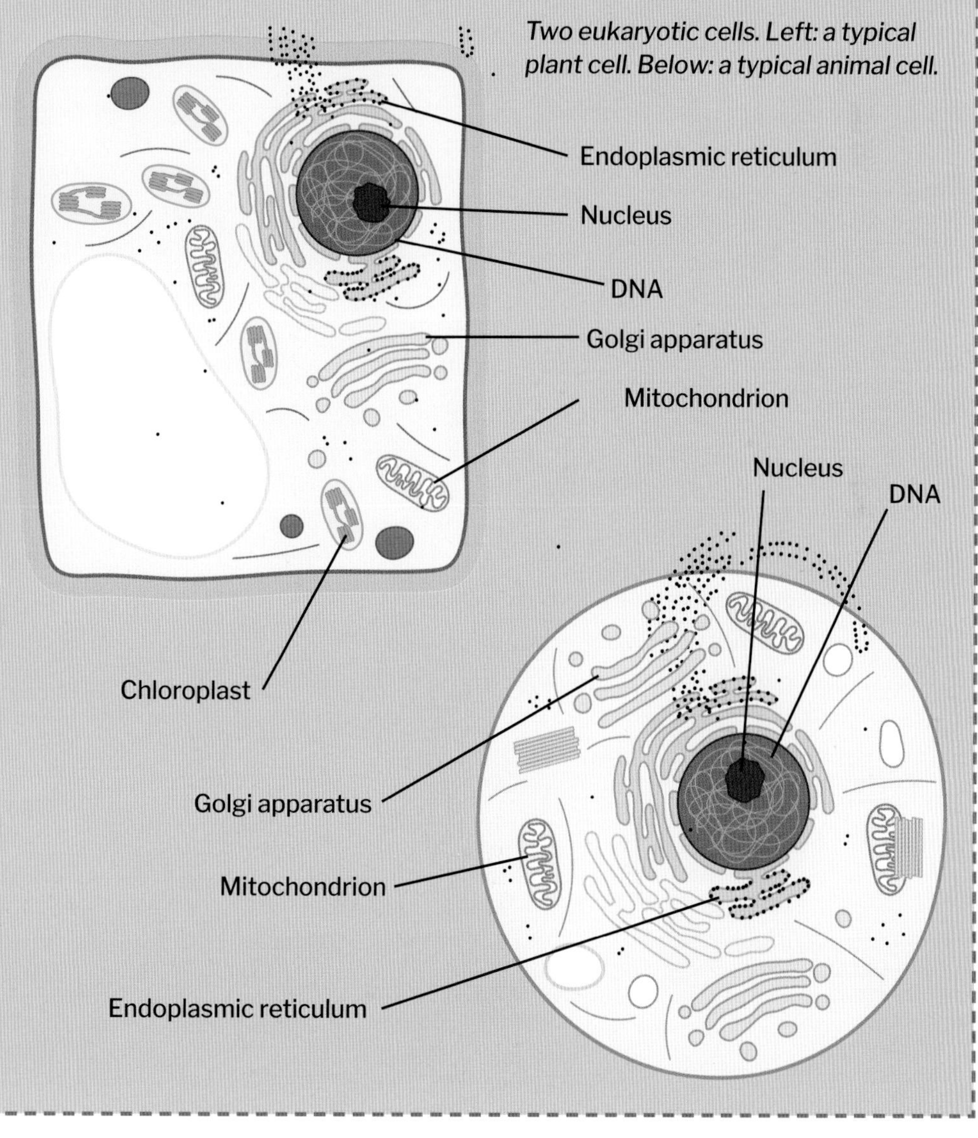

Two eukaryotic cells. Left: a typical plant cell. Below: a typical animal cell.

11

A central star

Heliocentrism

It seems obvious to us now that the Sun is at the middle of the solar system and the planets and other bodies orbit around it. Yet it was not always obvious, nor even the principal model of the solar system. For thousands of years, people believed Earth was at the middle, and the other planets, the Moon, and even the Sun went around Earth.

TWO OPTIONS

It's not clear just from observing the movements of the planets and the Sun which is the central object, Earth (a geocentric system) or Sun (a heliocentric system). The Ancient Greeks came up with both suggestions, but there was no way to make a definitive choice between them. The Earth-centred, or geocentric model, became prevalent. The influential philosopher Aristotle chose this model, and when the Greek-Egyptian astronomer Ptolemy wrote his influential astronomy text the *Almagest* in the 2nd century AD he put Earth at the centre. This ended further debate in the West for 1,500 years. Ptolemy described a complex arrangement of 'epicycles' and 'deferents' which had each planet not orbiting Earth directly, but orbiting an empty point which itself orbits Earth at a slight offset. This was necessary to explain the apparent backwards movements of the planets through the sky at certain points in their cycle. These backward, or 'retrograde', movements are the effect of observing the planets from Earth which is also in orbit. It has nothing to do with the actual orbit of the planets, which is always forwards.

Ptolemy's model of the universe.

GOD'S OWN UNIVERSE

The Ptolemaic view suited the Western religions since it put Earth, and so humankind, at the middle of the universe, just where we would expect God to put us. A verse in the Bible was even quoted in support of it, when God stops the movement of the Sun (rather than Earth) to prolong the day. With the support of the Church and Ptolemy, the geocentric model of the solar system went unchallenged for a long time. But that challenge came eventually.

Nicolaus Copernicus.

EARTH DETHRONED

In 1543, the Polish astronomer Nicolaus Copernicus published his account of the solar system, which put the Sun in the middle. This highly contentious view was published very shortly before his death, and the publisher (apparently under pressure from the Church) added a preface which suggested this was not a literal account of the solar system but was helpful in calculating the orbits of the planets. In fact, it didn't help, as Copernicus had all the planets in circular orbits around the Sun which didn't account for the apparent retrograde movement. He still needed the fudges set out by Ptolemy.

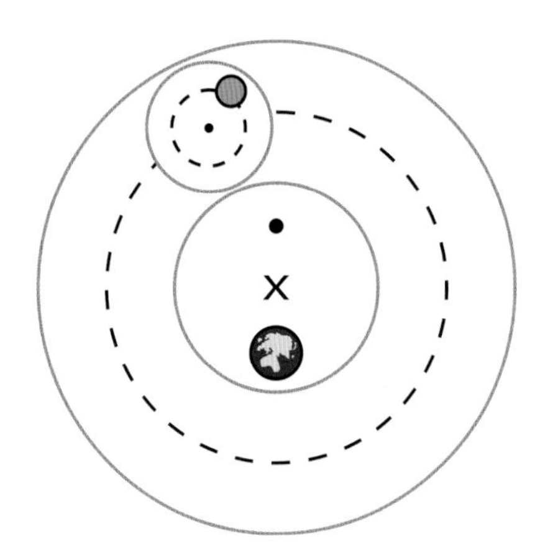

The planet goes around in a small circle (epicycle) and this circling system goes around Earth in an orbit called the 'deferent'.

At first, Copernicus's model drew little attention, but with the invention of the telescope in 1608, things began to change. The Italian polymath Galileo Galilei (1564–1642) discovered that the planets are other worlds, and some have their own moons in orbit around them. This made it harder to maintain that everything in the solar system orbits Earth. If the moons of Jupiter orbit Jupiter, there seemed less sense in assuming Jupiter then orbited Earth. When Galileo came out strongly in support of a heliocentric model of the solar system, the Church acted swiftly. He was questioned by the Inquisition (the Church's own, often savage, ecclesiastical court), placed under house arrest, and banned from teaching the heliocentric model.

ROUND AND NOT-SO-ROUND

Copernicus had made a crucial error in assuming that the orbits of the planets are circular. This was corrected by Johannes Kepler in 1604 when he realized the planetary orbits are elliptical – slightly squashed circles. The occasional retrograde movement of the planets was explained: Earth moves more quickly than the then-known outer planets (Mars, Jupiter and Saturn), so as Earth passes them in their orbits, they briefly seem to go backwards – just as a slow-moving car seems to go backwards as you pass it at speed. This made the heliocentric model clearly preferable.

FIGHTING A LOSING BATTLE

The Church continued to promote the geocentric model and tried to crush heliocentrism out of existence. Galileo's book that promoted the heliocentric model was put on the list of banned books in 1632 and not removed until 1822. Indeed, the Church didn't accept that its treatment of Galileo was unfair until the last decade of the 20th century. He's still waiting for his formal apology.

FINAL PROOF

Even though the mathematics of the heliocentric model works, and is supported by Newton's theory of gravity, it was still not possible to prove beyond doubt that it was correct until the 20th century. Only when we could put spacecraft into orbit to observe the movement of Earth and the other planets did it become absolutely clear that Earth orbits the Sun, and not the other way around.

The Copernican heliocentric model, with the known planets in orbit around the Sun.

12

In balance

Modelling the body

What keeps your body working and healthy? According to Ancient Greek theory, it was keeping the four 'humours' (bodily fluids) in balance. According to the even earlier Indian and Chinese thinkers, it was balancing energies and their flow through the body. If the body became unbalanced in either model, physical or mental illness was a likely result.

Health could be restored by nudging the body back towards its personal point of balance. In western medicine, the humoral model was championed by the influential Roman physician Galen and so prevailed for 2000 years. Balance has its place in the modern model of how the body works, too, but as part of a system that combines chemical and mechanical processes.

HUMOURS, SPIRITS AND ENERGY – NOT ENOUGH?

The various models of balance in the body began to face challenges from the Arab medical practitioners from around 900. Hints came that they might not be a complete explanation of health and illness. For example, Avenzoar (1091–1161) discovered that the skin condition scabies is caused by a parasite and not an imbalance of humours. But the little evidence that emerged was not enough to overthrow humoral theory. Around the world, religious and legal prohibitions prevented anatomists cutting into dead bodies, and that held back progress in understanding how the body works. Dissection of animal bodies gave some insights, but was also often misleading.

The turn towards a chemical and mechanical model of the body began with the rise of anatomy as a study based in dissection. The Flemish physician Andreas Vesalius presented the first real challenge to the work of Galen. Vesalius printed detailed anatomical drawings of the inside of the body and corrected several errors Galen had made and that had been perpetuated for 1,300 years. Now, at least, people had a reasonably accurate idea of what is inside the human body. That didn't rule out any of the balancing models – and indeed the humoral model clung on until the 19th century in many contexts.

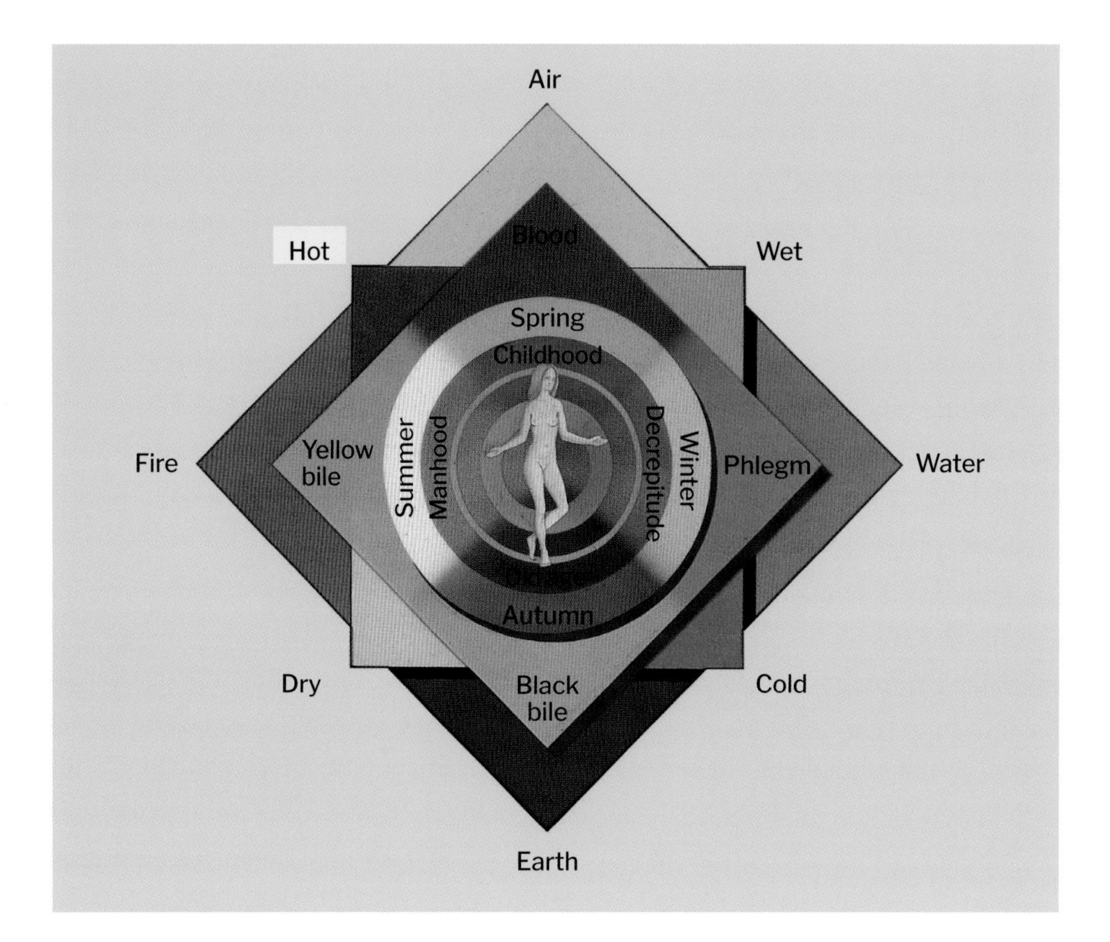

The model of four 'humours' or bodily fluids associated temperaments and ages with the characteristics associated with the humours. The fluids – blood, yellow bile, black bile and phlegm – were also linked with the four elements thought to make up all matter on Earth.

FROM FLUID DYNAMICS...

Identifying parts of the body didn't always show how they worked or what they did. That required a bit more work, often involving living bodies. The presence of blood vessels connecting all parts of the body to the heart and lungs was the first to be explained. The English physician William Harvey set out how the muscular heart pushes blood into the arteries with its regular contractions. He was not actually the first to notice this, but it was his account that had an impact on the accepted model of the body, finally discrediting the previous account of the blood ebbing and flowing through the blood vessels. Indeed, the Arab physician Ibn al-Nafis had already said in the 13th century that the blood picks up air from the lungs and returns to the heart.

Following work begun by al-Nafis and others, Harvey proposed that blood goes from the heart to the lungs to pick up air (we would now say only oxygen) and then travels around the body through the arteries, transferring to the veins through tiny vessels and so returning to the heart. The tiny connecting vessels were first seen by the Italian microscopist Marcello Malpighi in 1661. They are now called capillaries.

...TO MECHANICS

Blood pushed through tubes of decreasing width is subject to the laws of fluid mechanics. It soon emerged that other aspects of the physical body could be explained in terms of mechanical structures. The Italian scientist Galileo Galilei showed that muscles and bones work together in the same way as levers in a machine. The French philosopher René Descartes (1596–1650), inspired by seeing automata in Paris, thought the human body operated like a clockwork machine and that some actions are reflexes, caused only by mechanical processes and needing no thought. Descartes was far from ready to discard the spirit or soul, and the link between body and soul greatly preoccupied him. Even so, he explained involuntary reflexes, such as moving away from a flame, in terms of 'animal spirits' flowing though the nerves (then thought to be hollow) and exerting pressure that caused movement.

A CLOSER LOOK

The microscope, invented around 1600, soon provided medical science with a way of looking more closely at the structures of the body. Discoveries came thick and fast: cells of various kinds, sperm, bacteria, the fine structure of the alveoli (tiny sacs in the lungs), and others that did nothing to support the idea of humours. The discovery of capillaries and alveoli supported Harvey's explanation of the circulation of the blood. Isaac Newton explained the working of the eye in terms of optics. The magic of the body was being unpicked and explained in terms of ordinary phenomena. It made some people nervous.

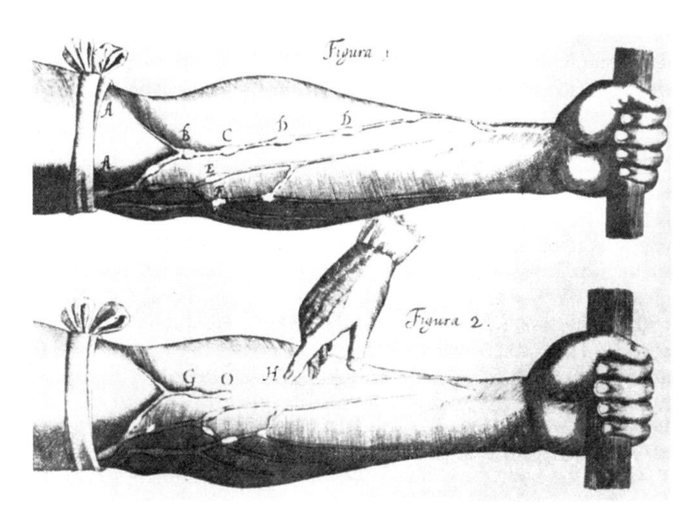

Harvey demonstrated that if blood flow is impeded, the veins swell – so blood is moving from the arteries to the veins for its return to the heart.

Spallanzani extracted part-digested food from unfortunate birds.

FROM MECHANICAL TO CHEMICAL

Although some body systems were obviously amenable to explanation following a mechanical model, digestion was a puzzle. Scientists argued about whether it was a purely mechanical process, with the churning of the gut breaking up the food, or whether digestion was essentially chemical. The other body systems were not seen in terms of chemical reactions, so this involved quite a new departure. Some rather unsavoury experiments settled the debate in favour of chemistry.

Francesco Redi (1626–98) carried out the first experiment, finding that in domestic fowl the gizzard can powerfully grind food – but humans don't have a gizzard. In 1752, René de Réaumur put morsels of food into a small, perforated metal tube which he attached to a string. He then induced a tame kite to swallow it. When he later removed it from the bird, he found the meat partly digested but the metal tube unharmed. The experiment was repeated and greatly extended by Lazzaro Spallanzani, who tried it with many different birds and animals. Like Réaumur, he found that gastric juice removed from the stomach could begin dissolving food.

Moving to human subjects, the American physician William Beaumont took advantage of an injury sustained by a young French-Canadian fur trapper in 1822. Beaumont's patient was badly wounded in the abdomen and when, against expectations, he recovered, he healed with a fistula, or hole, giving access to his stomach. Beaumont took advantage of this to experiment with digestion, poking foods in through the hole and removing them at different stages of digestion, as well as removing gastric juices to study. He found conclusively that while movement of the stomach muscles aids digestion, hydrochloric acid is mainly responsible.

NERVOUS ENERGY

The nerves presented more of a challenge. No one could find a fluid that ran through the nerves, but obviously they connected all parts of the body to the brain and spinal cord. Early theories favoured a 'balloonist' model in which the nerves carried fluid or gas that inflated muscles to contract them and produce movement. The role of electricity in nerve transmissions was discovered in 1771 by Luigi Galvani. He found that applying an electric current to the legs of frogs (removed from the frog) made them twitch as if alive. The nerves, it turned

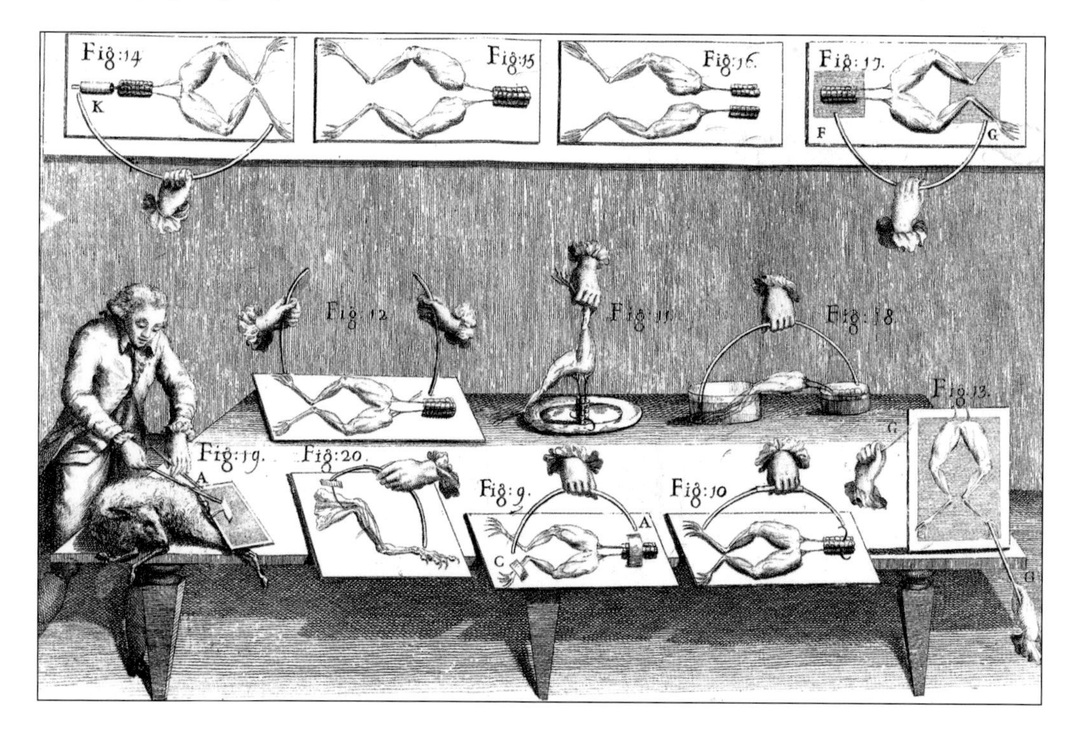

Galvani's frogs' legs experiment.

out, carried an electric current in the form of ions (charged atoms) passing electrons between them. The full method of the electrochemical working of the nerves was uncovered in the 1930s.

THE RETURN TO BALANCE

A modern take on maintaining the balance in a body is homeostasis. First suggested (though not named) in 1865 by Claude Bernard, it describes an organism's self-regulatory mechanisms for keeping conditions in its body optimal for life. That includes maintaining body temperature, pH, and a host of chemical balances. The body relies on sensing conditions, relaying that information to the brain, and then sending appropriate signals to correct any imbalance. This is mostly a matter of chemical balances, effected through the endocrine (hormonal) system. Its effects are often at the cellular level – it's very different from the supposed balancing of four fluid humours or of different types of energy in the body. It is part of a larger understanding of the highly complex mix of mechanical, chemical and bioelectric processes that keep the body working.

Claude Bernard teaching medical students.

13

Energy from atoms

Radioactivity

Without radioactivity, we wouldn't be here, yet no one knew it existed until 1896. In that year, French chemist Antoine Henri Becquerel had been working on naturally fluorescent material to study X-rays. He was trying to discover whether uranium salts could produce X-rays when, instead, he discovered the radioactivity of uranium. By the end of 1900 all three types of radioactivity were known: alpha, beta and gamma radiation.

A HAPPY ACCIDENT

Becquerel had been experimenting with exposing potassium uranyl sulphate crystals to sunlight, hoping to get them to absorb sunlight and emit X-rays. After their time in the sun, he planned to put them on a photographic plate to develop an image from the X-rays. But since the day was overcast, he put his crystals away in a drawer without any exposure to sunlight, wrapped in a dark cloth and alongside an undeveloped photographic plate. When he later developed the plate, he found a clear image, even though the salts hadn't had a chance to absorb any sunlight. He concluded that the uranium salts somehow emitted energy of their own that interacted with the silver bromide of the photographic plate.

Antoine Henri Becquerel.

MORE RADIATION

Marie Curie.

Polish chemist Marie Curie (Sklodowska) investigated uranium further and discovered that thorium produced the same kind of rays as uranium. She found that the level of radiation emitted by salts of the elements depended on the amount of the uranium or thorium present in them. From this she rightly deduced that the radiation was related to the atoms of uranium or thorium and not to the molecules they were involved in. It was a product of the element, not the compound. This was surprising, as the properties of compounds are usually quite independent of the properties of the elements that make up the compound. With her husband Pierre Curie, Marie went on to discover two new radioactive elements, which they named polonium and radium.

In 1898, Ernest Rutherford identified and named two distinct types of radiation, 'alpha rays' and 'beta rays'. The first he found to be a heavy particle with a positive charge and the second a much lighter particle with a negative charge. We now know that alpha radiation consists of streams of helium nuclei and beta radiation is streams of fast-moving electrons. The French chemist Paul Villard discovered the final type of radiation while working with radium. He

found that if he allowed radiation to pass through a thin sheet of lead, sufficient to stop alpha radiation, there were still two types of radiation present: one that could be deflected by a magnetic field and one that was unaffected by it. The latter was named gamma rays in 1903.

CHANGING ATOMS

An explanation of how these different types of radioactivity worked had to await understanding of the structure of the atom. We now refer to alpha and beta 'particles', and only gamma radiation is still called a 'ray'.

Alpha particles are the same as helium nuclei: two protons and two neutrons bound together. When an alpha particle leaves the nucleus of an atom of a radioactive element, such as uranium, the atom changes its atomic number, dropping by two – and this means it becomes a different element. Many radioactive elements go through a 'decay chain'. Their initial decay by losing an alpha particle produces another unstable atom that decays again into another element. For example, uranium (atomic number 92) decays into thorium (atomic number 90), which decays into radium (atomic number 88), and so on. Incidentally, the released alpha particles are the only natural source of helium on Earth, which is constantly replenished.

Beta particles are very fast-moving, high-energy electrons (or positrons, which are like an electron but with a positive charge). They are produced when a neutron in the nucleus decays into a proton, emitting an electron and an electron antineutrino.

Again, the number of protons in

Radioactive decay of uranium-234.

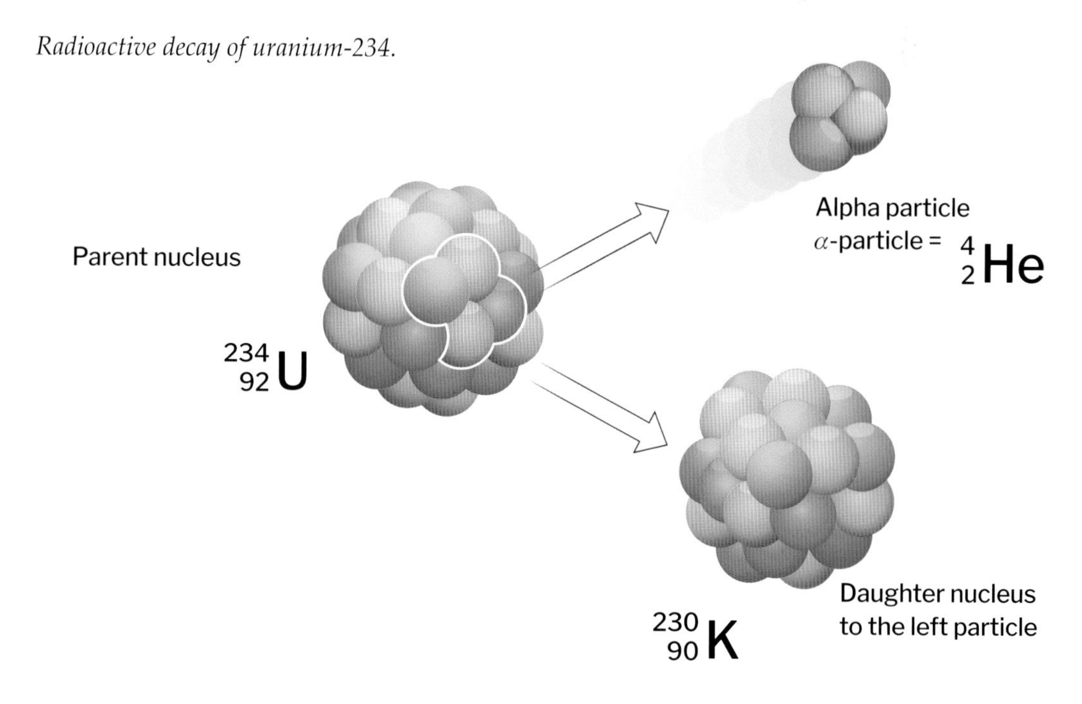

Radioactive decay of carbon-14.

the nucleus has changed, so the atom becomes a different element. When carbon-14 (atomic number 6) loses a beta particle it becomes nitrogen-14 (atomic number 7). Carbon-14 and nitrogen-14 are isotopes of carbon and nitrogen respectively. Although all atoms of an element must have the right number of protons, they can have different numbers of neutrons, which don't affect the overall charge of the atom. The total number of particles in the nucleus (protons plus neutrons) is the number shown for the isotope. So carbon-14 has 14 particles in the nucleus, which are six protons and eight neutrons. Nitrogen-14 still has 14 particles in the nucleus, but one neutron has switched to a proton, so now it has seven protons and seven neutrons.

Finally, gamma radiation is the emission of a high-energy photon. It is part of the electromagnetic spectrum, but the main source of gamma radiation is the radioactive decay of atomic nuclei. After alpha- or beta-particle decay, the atomic nucleus is in a high-energy 'excited' state and often emits energy in the form of a gamma ray photon.

Of the three types, alpha particles are the least penetrating, being stopped even by a sheet of paper. Beta particles can go further, but are still stopped by a thin sheet of metal. Gamma rays are the most penetrating and so the most dangerous to life. Typically, thick lead or concrete is needed to stop them.

GET A HALF-LIFE

Radioactive decay is a random process: it's impossible to predict exactly when an atom of a radioactive element will decay. Radioactive elements have different rates of decay and these are stated in terms of a 'half-life'. The half-life is the interval in which half of a sample of the element is likely to decay. Half-lifes vary enormously. The half-life of uranium-234 is 245,000 years, while that of the isotope (variant) uranium-238 is 4.5 billion years. On the other hand, thorium-233 has a half-life of 24 days, and radon-218 of just 0.0035 seconds.

14

Rock of ages

Ancient Earth

How old is Earth? It's a question that has been asked for millennia, but we have only been able to answer with any degree of accuracy in the last century. Some Ancient Greek philosophers believed that Earth was of infinite age; on the other hand, Christian thinkers of the 17th century considered it around only 6,000 years old, a date calculated by adding up the stated ages of Biblical figures from Creation to the birth of Christ.

FOSSILS HINT AT A PAST

Although people had found fossils for hundreds of years, they were often thought to have grown in the ground. In the 17th century, Danish naturalist Nicolas Steno first realized that there is a connection between the fossils present in a rock and the age of the rock. Recognizing the similarity between fossilized teeth and those of sharks caught in the Mediterranean, he proposed that fossils are the remains of once-living organisms. He also suggested that rocks have been laid down in layers over a very long period. The oldest rocks, having been laid down first, are at the bottom unless some event has disrupted them. Because fossils are from organisms that died when the surrounding rock was laid down, it's possible to give relative dates for fossils: the older fossils are buried below more recent fossils.

ANIMALS V. ROCKS

As more fossils were discovered, some of which looked nothing like modern organisms, the case for an ancient Earth grew more compelling. By the 19th century, physicists were arguing for an ancient Earth – but not ancient enough for some. In 1830, British geologist Charles Lyell argued that Earth is subject to steady but very slow change. Lyell was a great influence on the naturalist Charles Darwin. When Darwin outlined his theory of evolution, he argued for similar slow and steady changes in organisms. As there was no evidence of plants and animals having changed in recorded history, this suggested it would take a very long time for current organisms to have evolved, starting with the simplest single-celled organisms. Similarly, geological change is not obvious as we look around. The

Earth is actually over 4.5 billion years old.

biologists and geologists began to argue for a much older Earth than physicists were inclined to accept. But both argued for much longer than the 6,000 years the Church had favoured.

FROM HOT TO COLD

Physicists first tried to calculate the age of Earth from its supposed rate of cooling, assuming it began as molten rock. The first figure returned, by Georges-Louis Leclerc, Comte de Buffon in 1779, was 75,000 years. On the same basis, but differently calculated, William Thomson (later Lord Kelvin) put the age at somewhere between 24 and 400 million years in 1862. In 1897 he decided it was much nearer 20 million years old and no older than 40 million years. Kelvin assumed that initially Earth was

Charles Lyell.

uniformly hot all the way through and lost heat from the surface at a steady rate. In fact, Earth is hotter at the core and convection currents in the mantle and core mean heat moves around within the planet. But this wasn't the most important problem with his calculation.

Another way of calculating the age of Earth relied on working out how long it would take the oceans to accumulate the level of salt they have now. This was based on a (faulty) assumption that saltiness increased steadily. The calculation gave an age of 80–100 million years. In fact, minerals enter and leave seawater all the time, and the sea has had the same salinity for a very long time.

SUNNY DAYS

Kelvin and others also looked to the Sun in their attempt to work out the age of Earth. They felt that if they could work out how long the Sun would produce heat for, that would give an upper limit to the age of Earth. Unfortunately, they didn't know how the Sun produces its radiative energy, assuming that it was residual heat from the gravitational potential energy accumulated during the star's formation. A model that had the Sun gradually cooling, rather than actively producing energy, gave the Sun a lifespan of no more than 100 million years – so obviously Earth must be younger than that.

ON THE RIGHT TRACK

In 1895, John Perry (one of Kelvin's assistants) pointed out that Kelvin's assumption that Earth had a uniform temperature gradient, hot at the middle and cooling towards the outside, was not valid. When he used a model of a crust over a convecting mantle, Perry came up with an age of two to three billion years for Earth. This accorded much better with the age range that the geologists and biologists were arguing for, but Kelvin was having none of it.

More famously, the discovery of radiation provided a new potential source of heat for Earth. In 1903, the announcement that radioactive decay produces heat gave a more respectable challenge to Kelvin's estimate. He could hardly be faulted for not taking account of radiation when it was unknown. In fact, radiation makes little difference to Kelvin's calculation – it really needs convection currents as well to extend Earth's age much.

ROCK OF AGES, AGE OF ROCKS

The discovery of radioactivity was much more important to dating Earth in another way. As radioactive atoms in rocks decay they release an alpha particle (a helium nucleus) and so change the configuration of their nuclei, changing from one element to another. Ernest Rutherford, a physicist born in New Zealand but living in Britain, realized that by measuring the amount of a radioactive element still present, and the amount of helium trapped in a rock, and knowing the half-life of the radioactive element, it was possible to work out the age of the rock. He dated one rock he had to 40 million years, immediately establishing the lowest possible age of Earth. By 1907, American physicist Bertram Boltwood had dated rocks with ages ranging from 410 million to 2.2 billion years.

As radiometric dating has improved, the age attributed to Earth has increased to its present value of 4.55 billion years, ± 70 million years. This date has been established not from examining the rocks of Earth, but from meteoric rock formed at the same time as Earth in the early solar system.

Ernest Rutherford.

15

Making planets

Planetary accretion

Where Earth came from was treated by myth and religion for thousands of years before science came up with an explanation. Most myths involve gods or other supernatural beings making a fit home for humankind. The reality is rather different, with mankind entirely incidental and many other planets formed in the same way, though with different results.

FROM DUST

Earth and the other planets of the solar system formed from a giant cloud of gas, ice and dust whirling in a disk around the nascent Sun about 4.6 billion years ago. While the gas at the middle of the cloud was collapsing under increasing gravity to form a new star, tiny particles of solid matter orbited in a disk that stretched far out to space. This type of cloud has been seen around other new stars forming elsewhere in the universe and is called a 'protoplanetary disk'. The particles in the disk were specks of dust and grains of ice of different types. Some were debris from previous star systems that bequeathed their material to space; others had formed in the atmosphere of other stars and been pushed out into space.

All the dust, ice and gas circled the growing Sun in the same direction. As particles collided, they sometimes bounced off each other and sometimes stuck together, growing into large particles. As they grew, their gravity drew in more dust and grains, so they became ever larger in a process called 'accretion'. These lumps and clumps continued on their path, sweeping up the remaining dust and other chunks in their orbit, clearing their path through space.

AROUND AND ROUND

As the mass of a chunk increased with material added, its gravity pulled all of it inwards, towards the centre. With gravity acting equally on each part of the surface, the growing planets slowly became rounder and eventually nearly spherical. Each planet was going around the Sun but also rotating on its own axis. Each joining particle transferred mass and angular momentum to the growing clump of matter, keeping it spinning.

Formation of planets in a protoplanetary disc.

Initially, the axis of each planet would have been perpendicular to the plane of rotation of the protoplanetary disk, but some of the planets have been toppled or tilted by collisions. Earth's axis is at an angle of 22–24.5 degrees, knocked off the vertical by the collision that created the Moon. Uranus is lying on its side, its axis 98 degrees from the vertical. Venus is the only planet that spins in the 'wrong' direction. It might have been knocked so that it turned through 180 degrees, or perhaps it stopped spinning entirely and when it restarted it was going the other way.

ICE OR ROCK?

Our solar system has eight bodies large enough to be called planets. To qualify as a planet, a body must be large enough to have become round under its own gravity, orbit the Sun (not another planet), and have cleared its orbit of dust and other debris. Of those eight planets, the four closest to the Sun are rocky: Mercury, Venus, Earth and Mars. The outer four are made mostly of elements that on Earth are gases. The reason rocky planets lie closest to the Sun is that rock has a much higher melting (or freezing) point than any of the gases, and so even quite close to the Sun, rock particles are solid. Only beyond the frost line, between Mars and Jupiter, do the gases freeze into particles of ice. Ice particles can accrete into planets in the same way as rock particles. Although the rocky planets have some gas, which was originally clinging to the particles of rock and metal dust, they have a much lower proportion than the planets formed primarily from specks of various ices.

HOT IN THE MIDDLE

Gravity operates between objects with mass, and the greater the mass of an object the greater the gravitational attraction it exerts. As material was pulled inwards by gravity in the newly forming Earth, pressure increased at the middle, raising the temperature of the growing planet. Increased temperature allowed much of the planet to melt, and the material began to differentiate. The heaviest matter tended to sink towards the centre of Earth, leaving the lighter materials nearer the surface.

They came from outer space

Not all the matter in the protoplanetary disk formed into planets. Meteorites contain matter from the protoplanetary disk that accreted into lumps but didn't make it any further, never becoming part of a planet. They give us a good insight into the materials that Earth was originally made from. Unlike material on Earth, the rock of meteorites hasn't been subject to billions of years of change, weathering, and recycling, but is still close to its original form. Probes (robotic spacecraft) that visit other rocky planets and asteroids in the solar system can discover more about the very early history of Earth as these, too, formed from much the same materials as Earth.

The Hoba meteorite, which fell in Namibia, is the largest known intact meteorite on earth, weighing around 60 tonnes. It probably fell to Earth 80,000 years ago.

Iron particles melted and the heavy liquid trickled downwards between the particles of rock. Iron collected at the middle, forming the core, which is still mostly made of iron. This mass of iron had convection currents. The hottest iron, nearest the centre, rose as heavier cold iron fell inwards. Convection currents in the molten iron produced a magnetic field, which Earth still has.

MAKING A MOON

Accretion played out again in the formation of the Moon. Soon after Earth formed, just 60–200 million years after the formation of the solar system, Earth was in collision with another planet. Called Theia, and about the size of Mars, this planet crashed into Earth with such force that Theia was entirely destroyed and much of Earth was vaporized or hurled out into space. The vapour from the collision condensed in the cold of space. Some fell back to Earth, being reincorporated into the planet, but much remained in space, orbiting as small particles. These collided with one another and coalesced, accreting to form the Moon.

Theia collides with Earth.

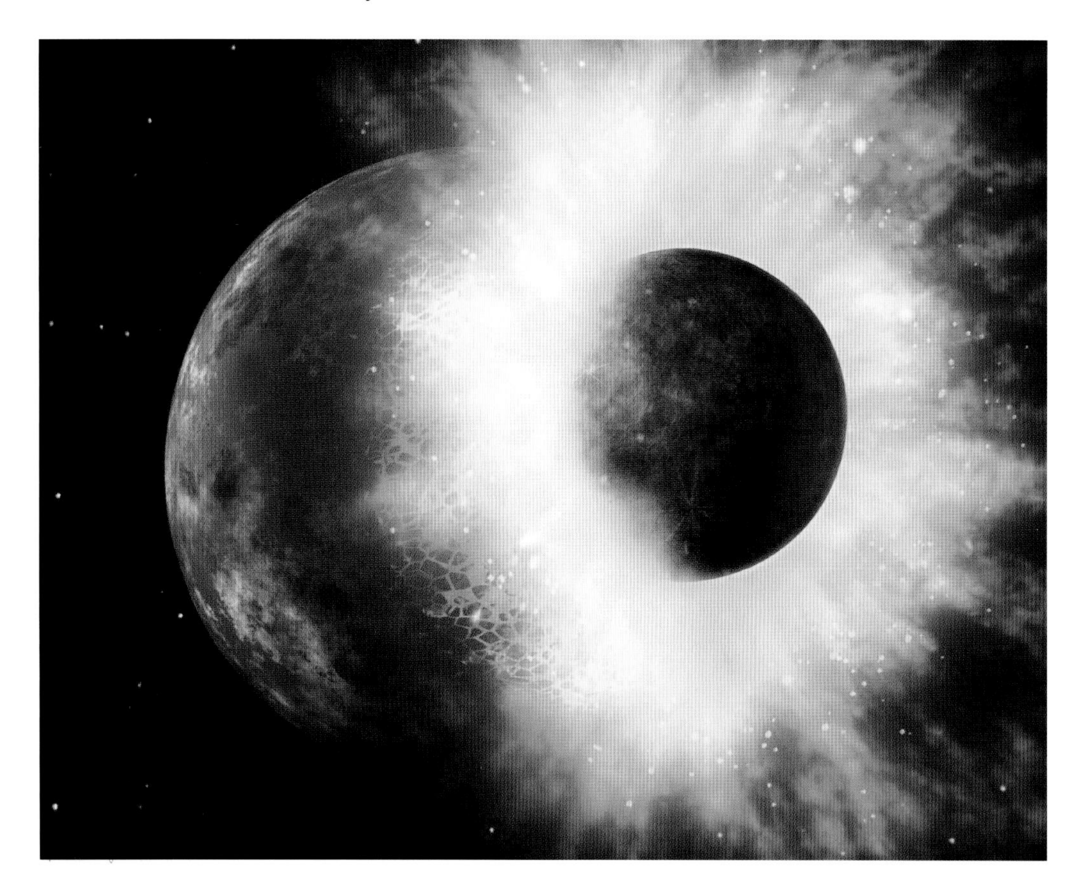

16

The smallest living things

Microbes

The idea that there are organisms too small to see with the naked eye is so familiar to us now that it's hard to imagine not knowing they exist. But if we step back to the invention of the microscope just before 1600, the concept would have been completely alien to most people.

There had been a few suggestions that there might be small things we can't see, possibly wreaking havoc. The Roman scholar Marcus Varro wrote 2,000 years ago in a book on agriculture that it was unwise to establish a farm near a swamp because 'there are bred certain minute creatures which cannot be seen by the eyes, which float in the air and enter the body through the mouth and nose and there cause serious diseases.'

INTO FOCUS

Micro-organisms ceased to be a subject of speculation and became real when first Robert Hooke and then Antonie van Leeuwenhoek saw them through their microscopes. Hooke was the first, publishing the first drawing of a micro-organism in his book *Micrographia* in 1665. It was the microfungus or mould *mucor*. Leeuwenhoek, who made microscopes that gave much greater magnification, found more types of microorganism a few years later. In 1676, he discovered

Antonie van Leeuwenhoek.

bacteria – and in considerable numbers. He saw so many in a drop of pepper-water that he wrote 'this exceeds belief'. But it was true, and confirmed the following year by Robert Hooke, to whom Leeuwenhoek had written about his discovery.

> *'Much to [my] wonder I discovered vast multitudes of these exceeding small creatures which Mr Leeuwenhoeck [sic] had described … and some of these so exceeding small that millions of millions might be contained in one drop of water.'*
>
> Robert Hooke, 1678

Leeuwenhoek saw not only bacteria but also protozoans and yeast – and hosts of other cells. Protozoans are single-celled organisms, many of which can move around. They include amoeba and *vorticella,* the first protozoan Leeuwenhoek described.

SOMETHING FROM NOTHING?

At first, people assumed that microbes came from nothing – that they spontaneously appeared in things like old food or compost. This notion, now labelled spontaneous generation, had organisms springing from inanimate matter in the right conditions. It had been suspected of larger things than microbes: worms, maggots, bees, and even scorpions and mice had been thought to appear spontaneously. Johannes Baptista van Helmont even published a recipe for mice in the 17th century: put some dirty rags or an old linen shirt into a barrel with some wheat or wheat bran, and 21 days later mice will have appeared in it!

> *'Your serpent of Egypt is bred now of your mud by the*
> *operation of your sun: so is your crocodile.'*
>
> William Shakespeare, *Antony and Cleopatra*, Act II, scene 7

Hooke's mucor.

The first person to challenge spontaneous generation rigorously was the Italian physician Francesco Redi in 1668. He set out an experiment with controlled variables – possibly the first such experiment – to determine whether flies appear in rotting meat spontaneously. He put fresh meat into three sets of wide-necked jars. One set he left open to the air. One set he closed tightly with lids. He covered the openings of the final set with gauze. Then he watched and waited. Flies landed on the open jars, and later maggots appeared, and after that he saw more flies around that meat. In the sealed jars, no flies could visit the meat, and no maggots or flies appeared in the jars. In the final set, flies laid eggs on the gauze but no maggots could get access to the meat, so no flies appeared in the jars. This was taken as conclusive proof that flies only come from other flies – they don't appear directly from meat.

Flies are relatively large and complex, at least when compared with microbes. It still remained plausible, and even likely, to most people that microbes could spontaneously generate from inanimate matter. It was commonly thought that a 'life force' was present in all inorganic matter, including air, and this could be held responsible for making animate organisms from inanimate matter.

SOUP AND MORE SOUP

In 1745–8, Irish clergyman Joseph Needham found that microbes grew in broth exposed to the air, and even in broth that had been briefly boiled and put into a flask sealed with a cork. Twenty years later, Italian biologist Lazzaro Spallanzani (he of the force-fed kite, see page 56), carried out a more rigorous experiment. He boiled flasks of broth for an hour and then sealed them by melting the glass at the opening, shutting the sides together. He boiled some flasks for

Francesco Redi's experiments.

Open jar

Gauze-covered jar

Sealed jar

just a few minutes and sealed those. And then he boiled some flasks for an hour and closed them with a cork stopper. The flasks that had been boiled for just a few minutes soon grew microbes, but those boiled for an hour stayed clear. Needham's answer was that Spallanzani had killed the 'life force' by prolonged boiling and sealing the flasks meant no new life force could enter.

The case was finally solved by Louis Pasteur in 1864. Responding to the offer of a prize by the Paris Academy of Sciences, he devised a method of allowing air – but not microbes – into flasks of boiled broth. This would allow any 'life force' to act, if it existed. He boiled liquid in swan-necked flasks. He then broke the necks off some flasks, allowing air in. The liquid in these flasks soon spoiled. Others, he left with the swan-neck intact or he plugged with cotton. In these, the liquid remained pure: microbes were caught by the cloth or could not rise up again after the lowest bend in the swan-neck, collecting at the bottom. Here, at last, was proof that not even microbes generate spontaneously.

'Never will the doctrine of spontaneous generation recover from the mortal blow of this simple experiment. There is no known circumstance in which it can be confirmed that microscopic beings came into the world without germs, without parents similar to themselves.'

Louis Pasteur

The design of Pasteur's swan-necked flasks prevented microbes reaching the contents.

17

Spreading diseases

Germ theory

Illnesses can be divided into those caused by pathogens (germs) and those caused by the body going wrong in some way. A condition like cancer or heart disease is not something you can catch from another person, but infectious diseases such as flu, measles and polio, are caused by germs that can be passed between people. But this is quite a recent insight.

BAD AIR AND BAD HUMOURS

An early model of health and sickness had the body composed of four humours (in the West) or balanced forms of energy (in the East). When these were out of balance, the result was illness. The means of cure was thought to be restoring balance. But while it was easy to see how this might explain individual problems, it didn't make much sense in the case of an epidemic. Contagion was often explained as the result of 'bad air', or 'miasma' – malaria even means 'bad air'. Bad air was supposedly produced in damp areas by rotting vegetation and other matter. The foul-smelling air was thought to be filled with tiny particles of decayed matter (miasmata) which produced illness. As people living close together shared the same air, this explained why many fell ill at the same time.

As cities grew in the Industrial Revolution, people often lived in cramped and unhygienic conditions. The River Thames in London and the Seine in Paris were filled with rotting refuse, and clusters of disease near the rivers seemed to support the miasma model. Diseases such as cholera spread quickly and were often blamed on the low air quality. The air really was of low quality in 19th-century cities, and caused many health problems, but it wasn't responsible for cholera which is caused by bacteria, usually taken in through contaminated drinking water. Measures taken against the supposed miasma often did reduce the incidence of disease as they involved better sanitation and hygiene measures. These reduced the load of bacteria and viruses and so cut infections – but that reinforced the idea that disease was caused by miasma.

Pacini's slide identifying cholera.

FIGHTING CHOLERA

The miasma model was challenged by the discovery by John Snow in London in 1854 that a cholera outbreak could be traced to a particular water pump. Water from the pump was contaminated with sewage, and people drinking the water from it were taking in cholera bacteria. Closing off the pump reduced cases of cholera. The same year, the Italian anatomist Filippo Pacini discovered the cholera bacillus that causes the disease – but the miasma model was so firmly rooted that his discovery had no impact. Pacini carried out autopsies on patients who died of cholera during an epidemic in Florence, Italy, and found the microscopic, comma-shaped bacterium in the lining of their guts.

Filippo Pacini.

KOCH'S THEORY OF GERMS

German microbiologist Robert Koch firmly established the link between a bacterium and a disease in 1890, working with anthrax. He discovered and isolated rod-shaped bacteria in the blood of cows suffering from anthrax and infected mice with them. When the mice developed anthrax, he could confidently claim that he had found the cause of a particular disease. He came up with four statements, known as Koch's Postulates, which establish a causal connection between a bacterium and a disease:

- The bacterium is present in every case of the disease, but not in healthy organisms.
- The bacteria isolated from a host with the disease are grown in pure culture.
- A healthy host given the bacteria develops the same disease.
- The same bacteria can be recovered from the host that has been infected.

Koch identified the bacteria that caused tuberculosis, anthrax and cholera (rediscovered after the world had ignored Pacini's discovery).

BEYOND BACTERIA

While Koch was able to isolate and identify bacteria, and establish their role

Koch's postulates

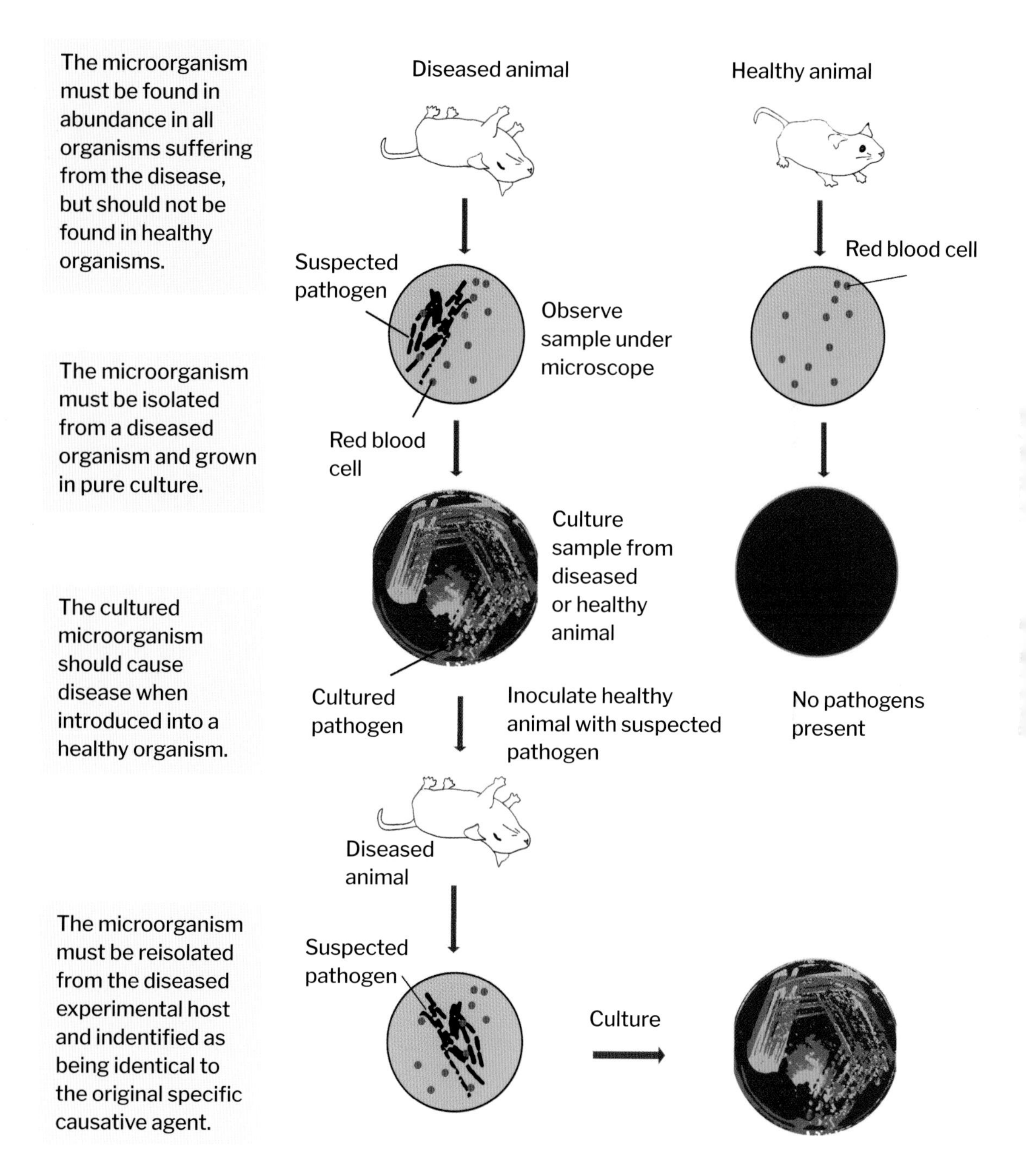

Robert Koch.

as pathogens, some infectious diseases eluded exploration in this way. Rabies, which Louis Pasteur studied (and produced a vaccine for) is one example. It proved impossible to isolate a bacterium, and Pasteur pondered the possibility of a smaller agent of disease than bacteria. Rabies is caused by a virus, and viruses are much smaller than bacteria – too small to be seen with the microscopes available to Pasteur and Koch in the 19th century.

In 1892, the Russian microbiologist Dmitry Ivanovsky explored tobacco mosaic disease, which affects tobacco plants. He found that if he took sap from infected leaves and passed it through a porcelain filter, the liquid was able to infect healthy leaves. A porcelain filter has such fine holes that bacteria can't pass through, so his experiment demonstrated that the disease is not caused by a bacterium. It doesn't seem that Ivanovsky fully grasped what he was dealing with, and suggested the disease might be caused by a toxin the bacteria produced.

Further investigation was left for the Dutch microbiologist Martinus Beijerinck to discover. In 1898, he referred to what he had found as a 'contagium vivum fluidum' or a living germ soluble in liquid that could cause contagion. Remarkably, it reproduced in close association with the metabolism of the infected host organism. This didn't fit with germ theory and so was not immediately popular. We now know that viruses can only reproduce in a host's cells. They hijack the cell's reproductive mechanism and resources to make copies of themselves. The viral particles then leave or burst out of the cell to infect further cells.

Viruses don't behave in the same way as bacteria, being unable to reproduce outside a host, so didn't fit with Koch's postulates – they couldn't be grown in a culture medium before being used to infect a new host. Still, the virus-infected liquid did produce the same disease in a new host, and further contagious liquid could be extracted from the new host. Beijerinck maintained viruses were liquid, but in 1898 the American biochemist Wendell Meredith Stanley showed they are particles. Particles of tobacco mosaic virus were finally isolated in 1935.

18

Shining a light on light

People began to think about the nature of light a long time ago. The questions that preoccupied early thinkers and scientists working on optics were how vision happens, whether the speed of light is finite or infinite, and the nature of light (whether it's rays or particles).

LOOKING INTO LIGHT

The notion that light involves some type of ray that travels in straight lines was proposed more than 2,000 years ago and formed the core of work by the Ancient Greek mathematicians Euclid (*c.* 325–265 BC) and Hero of Alexander (*c.* AD 10–70). Hero worked out that the image in a mirror seems to be as far behind the mirror as the object reflected is in front of it. While this was correct, his explanation of vision as produced by rays emitted from the eye bouncing off objects in the world was wrong. This – the 'extramission' theory of vision – prevailed for hundreds of years.

Arab scientists took up the baton from the 9th century AD. Al-Kindi (*c.* 801–873) suggested that 'everything in the world ... emits rays in every direction, which fill the whole world.' Ibn al-Haytham, in the early 11th century, wrote extensively on optics, following his theory that light and colour are formed by rays that can reflect, refract and so on. He argued that vision is produced by rays coming into the eye, not emitted from it, and that the speed of light is finite, though very fast. Arab work on light and lenses laid the foundation for two of the most important scientific inventions of all time – the microscope and the telescope, both developed around 1600.

In Europe, Isaac Newton (1642–1727) was the master of light. He showed

Ibn al-Haytham.

that white light can be broken into a spectrum of coloured light by a lens and reconstituted into white light again with no change to its qualities. He argued that the colours we see in the world around us are produced by objects interacting with coloured light, rather than objects generating the colours. Newton claimed that light is composed of particles, which he called 'corpuscles'.

WAVES?

Early thinkers had treated light as rays when working on refraction and reflection. While Newton thought it consisted of a stream of high-speed particles, his Dutch contemporary Christiaan Huygens argued that light propagates as waves. The debate seemed to be settled by Thomas Young in 1803 with a famous experiment that involved passing light through two slits. The resulting interference patterns showed light behaving in just the same way as waves of water passing through two openings, the two sets of waves interacting to reinforce or cancel each other in a characteristic pattern.

It was as a wave that light entered James Clerk Maxwell's electromagnetic spectrum (see page 84). The colours that Newton had found in white light could now be divided by wavelength. The white light of sunlight, it emerged, is a mixture of light of different wavelengths and is combined with other wavelengths also coming from the Sun, across the full electromagnetic spectrum.

OR PARTICLES?

All seemed well with light as waves until 1899. That year Max Planck, working on black-body radiation, suggested that light is only released in tiny, discrete quantities. Black-body radiation explains how the light emitted by an object changes

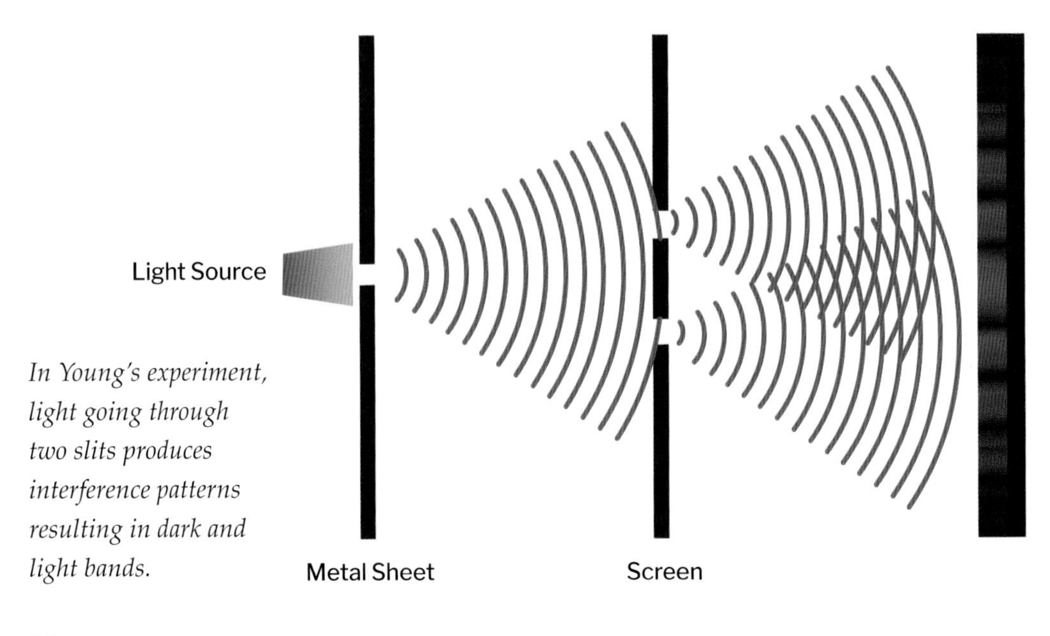

In Young's experiment, light going through two slits produces interference patterns resulting in dark and light bands.

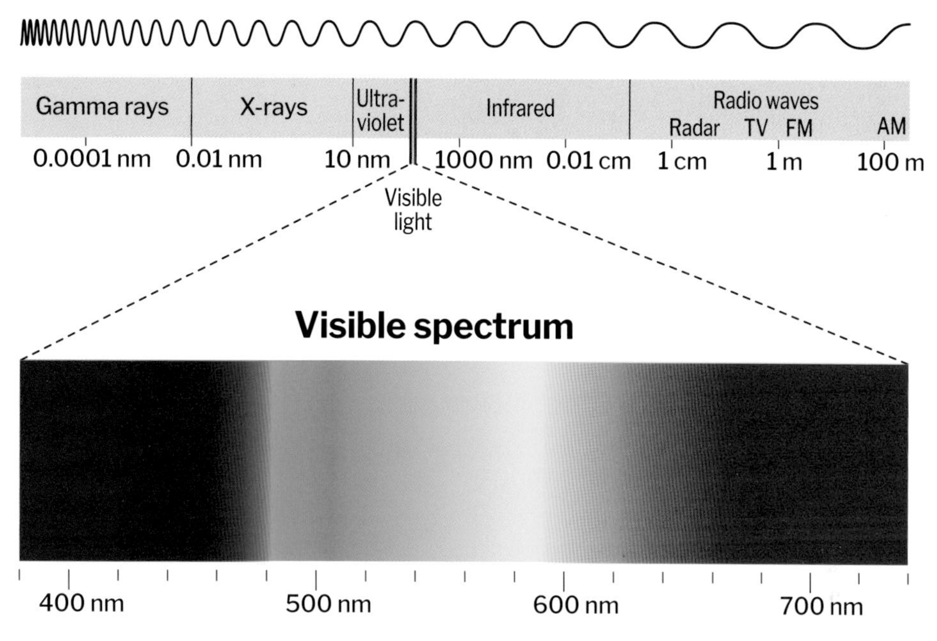

Visible light spectrum.

with its temperature. A cold object emits only long-wave length radiation that we can't see, but as it gets hotter it begins to emit red light, then orange, yellow and eventually bluish-white. A 'black body' is a hypothetical object that absorbs all the radiation that falls on it, though nothing is truly that black.

The speed of light

The question of whether light travels at a finite speed or spreads everywhere instantaneously has been a vexing one since the time of the Ancient Greeks. Galileo is credited with the first attempt to measure the speed of light in 1638. His method relied on working with an assistant. Galileo and the assistant each held a lantern at a distance from the other. One uncovered his lantern first, and the other uncovered his lantern as soon as he saw the light. Needless to say, the slowest part of this was their reaction times. Galileo concluded that light travels at least ten times as fast as sound. Ole Römer first measured the speed of light in 1675 at about 200,000 km (125,000 miles) per second. In 1738, English physicist James Bradley measured the speed of light at 301,000 km (187,000 miles) per second, close to the actual speed of 299,792 km (186,000 miles) per second.

The photoelectric effect: light rays knock electrons out of a layer of metal.

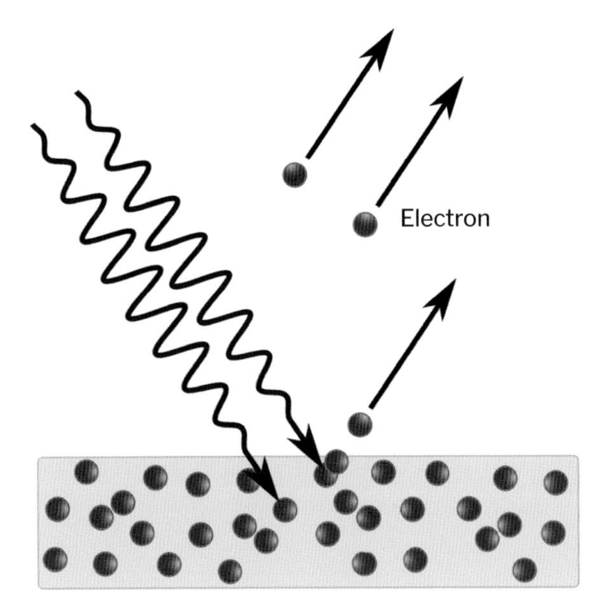

Planck named the tiny packets of energy 'quanta'. It wasn't at first clear whether the quanta were a feature of light itself or of its interaction with matter – that is, whether light is innately divided into quanta or whether matter breaks it into quanta. That question was answered by Albert Einstein in 1905 in his investigation of the photoelectric effect. Light, it transpires, is innately divided into quanta, which we now call photons. Photons of light of different wavelengths have different energy.

OR BOTH?

It turns out that light behaves as both a particle and a wave at the same time. Sometimes, depending on how it's investigated, it seems to act more like a wave. At other times, particle-like behaviour is more obvious. Photons are odd particles, though. They have no mass or energy at rest. They move at the speed of light in a vacuum. They can interact with other particles, despite having no mass. They can be instantly created or destroyed (in radioactive decay, for example). And their energy and momentum depend on the frequency and wavelength of the electromagnetic wave they are part of.

Electricity from light

The photoelectric effect is central to solar power generation, so is increasingly important in our world. It rests on the ability of photons to interact with other particles. In particular, the 'Compton effect': photons colliding with an atom can knock an electron free from the atom. The stream of electrons knocked out of a metal or semiconductor (such as silicon) is captured to produce an electric current. When the photons are coming from the Sun, that's essentially a free source of energy in that it doesn't consume anything in generating electricity.

19

From radio to gamma rays

The electromagnetic spectrum

It's far from obvious that the radio waves which carry your mobile phone signal are the same type of phenomenon as the microwaves which heat up your dinner, the X-rays that reveal dental problems, and the light that shines from your spotlights. Yet all are forms of electromagnetic radiation (EMR).

They fall in different parts of the electromagnetic (EM) spectrum, with different wavelengths producing energy that has different properties and that we experience differently.

WAVES OF ENERGY

The EM spectrum ranges from the very long wavelength of radio waves to the very short wavelength of gamma-ray radiation. The wavelength is the distance between the peak of one wave and the peak of the next. For radio waves, the wavelength is measured in metres – up to thousands of metres. Microwaves fall between many centimetres and a

The electromagnetic spectrum.

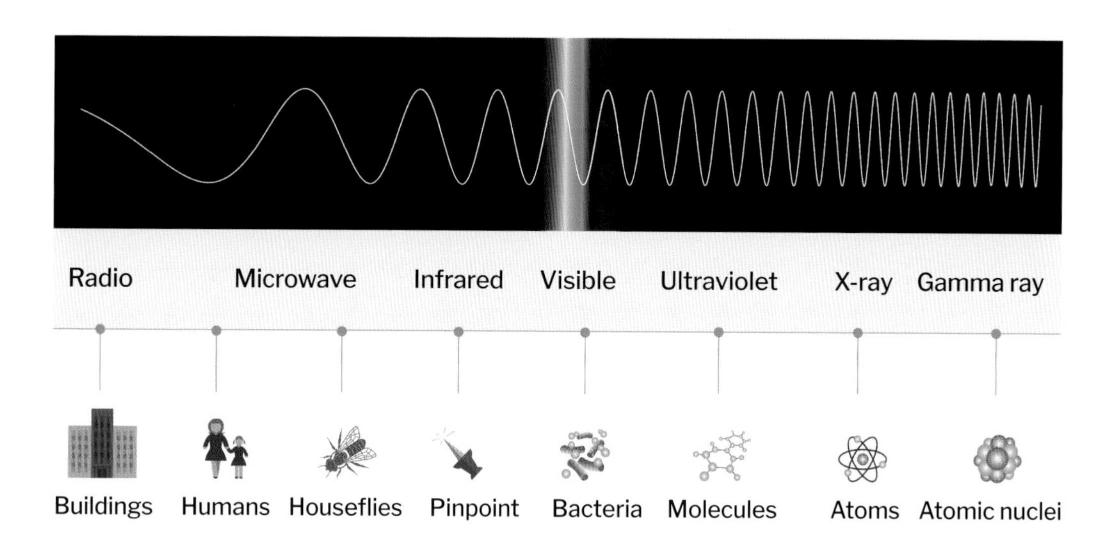

millimetre (a thousandth of a metre), and that of gamma-ray radiation is measured in trillionths of a metre (billionths of a millimetre), or about the scale of atomic nuclei.

As well as a wavelength, waves of energy have a frequency. This is the number of waves that pass a given point in a specified interval. Frequency is measured in Hertz (1 Hz = one cycle per second). The smaller the wavelength of radiation, the higher the frequency and the more energy transmitted. Because of the very high energy levels of X-rays and gamma-ray radiation, these are very harmful to our bodies. Radio waves are harmless, having a very long wavelength and low level of energy.

James Clerk Maxwell.

FROM ELECTRICITY AND MAGNETISM

The Scottish physicist James Clerk Maxwell (1831–79), probably the greatest physicist of the 19th century, found that electricity and magnetism are not entirely different things, as they at first appear. Instead, they are different manifestations of the same force – the electromagnetic force. Maxwell predicted that waves of this force are propagated and spread outwards, like ripples in a pond, at the speed of light. He concluded that visible light is a form of electromagnetic wave, consisting of waves of electrical field and waves of magnetic field that vibrate perpendicular to each other and to the direction in which they are moving.

His identification of light as a form of EM radiation, and his move from talking about electromagnetic forces to electromagnetic fields, were the most important developments in physics since the work of Newton. He published his theory of electromagnetic radiation in 1862 and 1864, paving the way for Einstein's work on relativity and the development of quantum physics.

WAVES THROUGH WHAT?

We're used to thinking of waves passing through a medium. Waves at the beach are carried by water and sound waves travel through air (or other substances – but they can't travel through empty space). It was natural to look for a medium to carry electromagnetic waves, and the solution proposed was the 'aether', a hypothetical substance that fills all

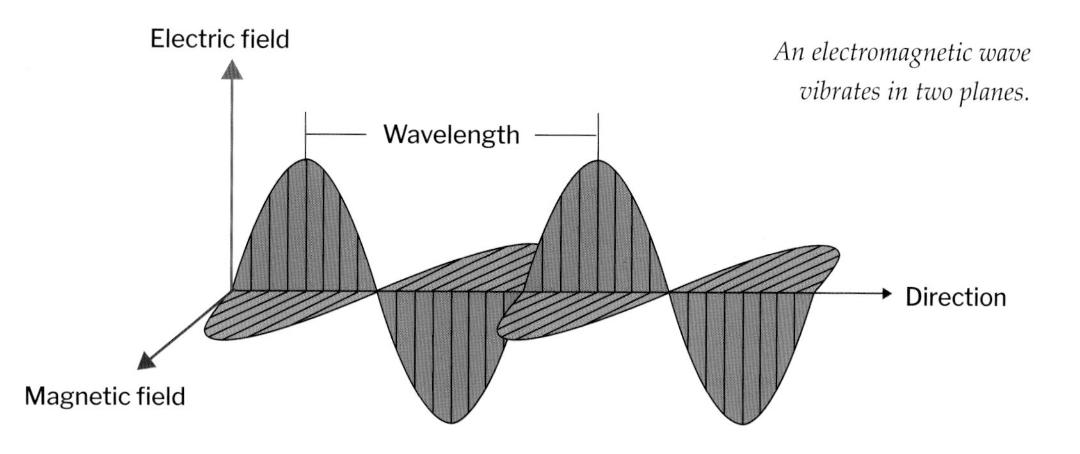

An electromagnetic wave vibrates in two planes.

space. At first, Maxwell also assumed that electromagnetic waves should flow through aether, and it remained a potent and popular idea for a few more years.

The existence of the aether was finally disproven in a famous experiment carried out by American physicists Albert Michelson and Edward Morley in 1881 and again in 1887. Michelson and Morley had set out to measure the speed of light in two directions in the expectation of showing that it was faster when moving with the flow of aether than across it. This relied on Earth moving through space producing an 'aether wind'. When they could find no difference in the speed of light in different directions, the existence of the aether began to look highly doubtful. The Michelson–Morley experiment has been considered one of the most important 'failed' experiments.

MORE AND MORE WAVES

Infrared had been discovered by astronomer William Herschel in 1800 and ultraviolet by Johann Ritter the year after. Heinrich Hertz produced and studied microwaves in 1886 and first transmitted radio waves in 1888, confirming Maxwell's hypothesis that there could be more types of ER to discover. Wilhelm Röntgen discovered X-rays in 1895, but they weren't identified as part of the electromagnetic spectrum until 1912. In 1897, Henri Becquerel and Ernest Rutherford discovered three types of radiation: alpha, beta and gamma rays. The first two turned out to be particles (helium nuclei and high-speed electrons respectively), but gamma rays were identified as EM radiation in 1912.

PUT TO USE

When radio waves were discovered, it was generally thought they wouldn't be much use. Now we use them in most aspects of our lives as they transmit the signals we use for radio, television, mobile phones and the internet. Radio waves can be made to carry information by varying the amplitude of the waves. This is the height of each wave. The wavelength

From Aristotle to Einstein

'Aether' has its origins in Greek mythology as a particularly pure essence breathed by the gods. The philosopher Plato then referred to it as an extra-pure type of air beyond the four traditional elements: earth, water, fire and air. His pupil Aristotle proposed an extra element which would fill space beyond Earth. He didn't call it aether, but the name was soon adopted by later thinkers. Newton saw the aether as a medium for the transmission of gravity. When physicists began thinking in the 17th century about waves of light propagating through space, the aether was the natural medium for them. It even became called the 'luminiferous aether'. It was finally rendered redundant by Einstein's theory of special relativity.

and frequency remain the same, but by varying the height of the wave we can encode audio or other information into them.

Radio waves and other parts of the EM spectrum have other uses too. Radio telescopes help us map the skies and examine astronomical objects that don't emit visible light. Telescopes can also use microwaves, infrared and ultraviolet in the same sort of way, giving us new insights into objects such as pulsars and distant stars that emit radiation across the spectrum.

We use microwaves to heat food by harnessing the energy of the waves to vibrate molecules in the food. X-rays let use see 'through' many types of matter, allowing us to look inside bodies for medical purposes and into objects, such as luggage and ancient artefacts. X-rays are absorbed and reflected in different ways by different types of matter. They can also be used to penetrate the body and irradiate cancer cells, using the energy of the waves to destroy the targeted cells. Gamma rays can also treat cancer, and new 'gamma knife' technology uses a targeted beam of gamma rays for very precise brain surgery.

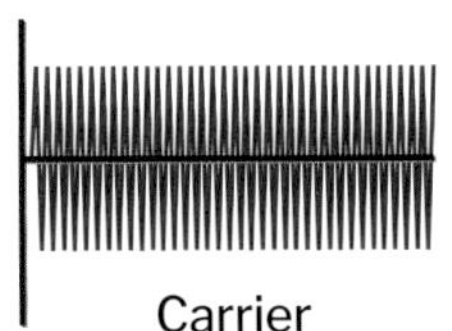

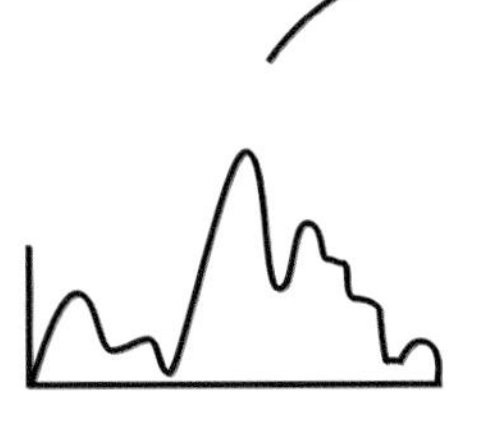

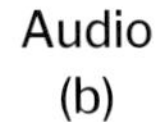

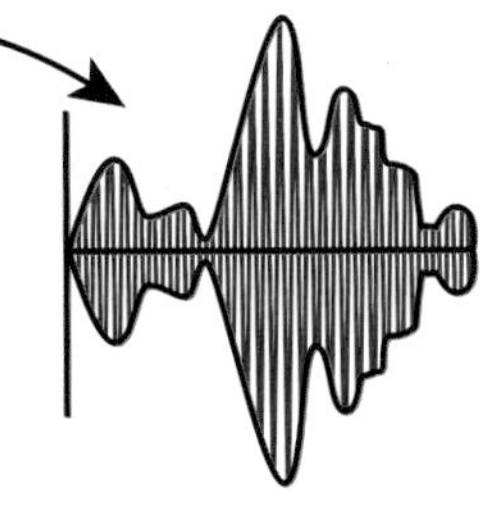

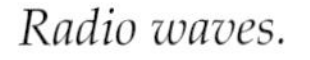
Radio waves.

20

The origins of life on Earth

Abiogenesis

Earth is teeming with life, from microbes only visible with a microscope to giant trees and blue whales. Where did it all come from?

We don't know exactly how many species there are. One fairly recent estimate suggests 8.7 million – but it could be up to a trillion, most of which would be microbes. These are the product of billions of years of evolution, but it all started with some type or types of simple microbe. There's a lot we don't know about the first life on Earth – including when, where, how and why it began. Somehow, something made the jump from inanimate, unliving matter to something living. That process is called 'abiogenesis'.

BIG QUESTIONS: WHAT IS LIFE?

Untangling how life started isn't helped by us not having a thorough and universally agreed definition of 'life'. A fairly basic definition says a living thing must be capable of reproducing itself and of taking energy from its environment to grow and reproduce. It's obvious something like a mushroom or a zebra does this.

Among the earliest known lifeforms are microorganisms that have produced stromatolites, rocky humphs of fossilized microbes laid down in layers.

Even microbes are clearly living by this definition. But what about a virus? Is it living or not? It can't reproduce on its own – it has to take over the machinery of a host cell and force that to make copies of it. Many scientists consider viruses to be on the boundary between living and non-living things. That boundary is an interesting place, since at some point our biochemical ancestors must have crossed it.

BIOLOGY FROM CHEMISTRY

There are several suggestions as to how life started. All possibilities need some source of energy to kick-start a reaction that would produce chemicals capable of reproducing themselves. Self-replicating chemicals are at the heart of all living things. DNA copies itself every time cells divide. The energy to set this process going could have come from lightning striking a pool of water, from the geothermal heat of a deep-sea vent, or perhaps just from sunlight.

SPARK OF LIFE?

In a famous experiment in 1953, chemists Harold Urey and Stanley Miller tried to reproduce conditions on early Earth to see if they could make amino acids (see box on page 90) from the chemicals they thought were available in Earth's atmosphere at the time. They made a mix of gases to mimic the early atmosphere and zapped it with electricity like lightning. They did succeed in making amino acids – more than they realized, in fact, as scientists looking at the samples 50 years later found more than Urey and Miller had recorded. In a second experiment in 1958 they used a different combination of gases including hydrogen sulphide and carbon dioxide. This combination could have been found around active volcanoes. It produced more chemicals associated with life: 23 amino acids and four amines. The experiment doesn't prove that amino acids were first created like this, but it does demonstrate that they could be created from gases in the atmosphere.

MADE FROM CLAY

Another idea is that replication began with organic molecules sticking to clay crystals. Crystal structures show inorganic chemicals organizing themselves in a consistent pattern. It's possible that organic chemicals effectively used clay as a reproduction medium, later breaking free to repeat their sequences on their own.

TRAPPED INSIDE

However the first self-replicating chemicals came about, to make something living they had to be more than just chemicals in the early waters or atmosphere. One possibility is that some became trapped inside capsules of a special type of molecule that has a water-hating (hydrophobic) end and a water-loving (hydrophilic) end. These molecules naturally group into spheres in water as the hydrophobic ends all hide inside while the hydrophilic ends are in contact with the water. If they form a double layer, they make a wall that can contain a trapped drop of water. The hydrophobic tails are in the middle of the wall while the hydrophilic heads are in the water inside or outside the capsule.

If the trapped water contained a self-replicating chemical, it would form a prototype for a very simple type of cell. The insides are separated from the outside water, the capsule can grow by adding to its wall and by the chemicals

The equipment used in the Miller–Urey experiment.

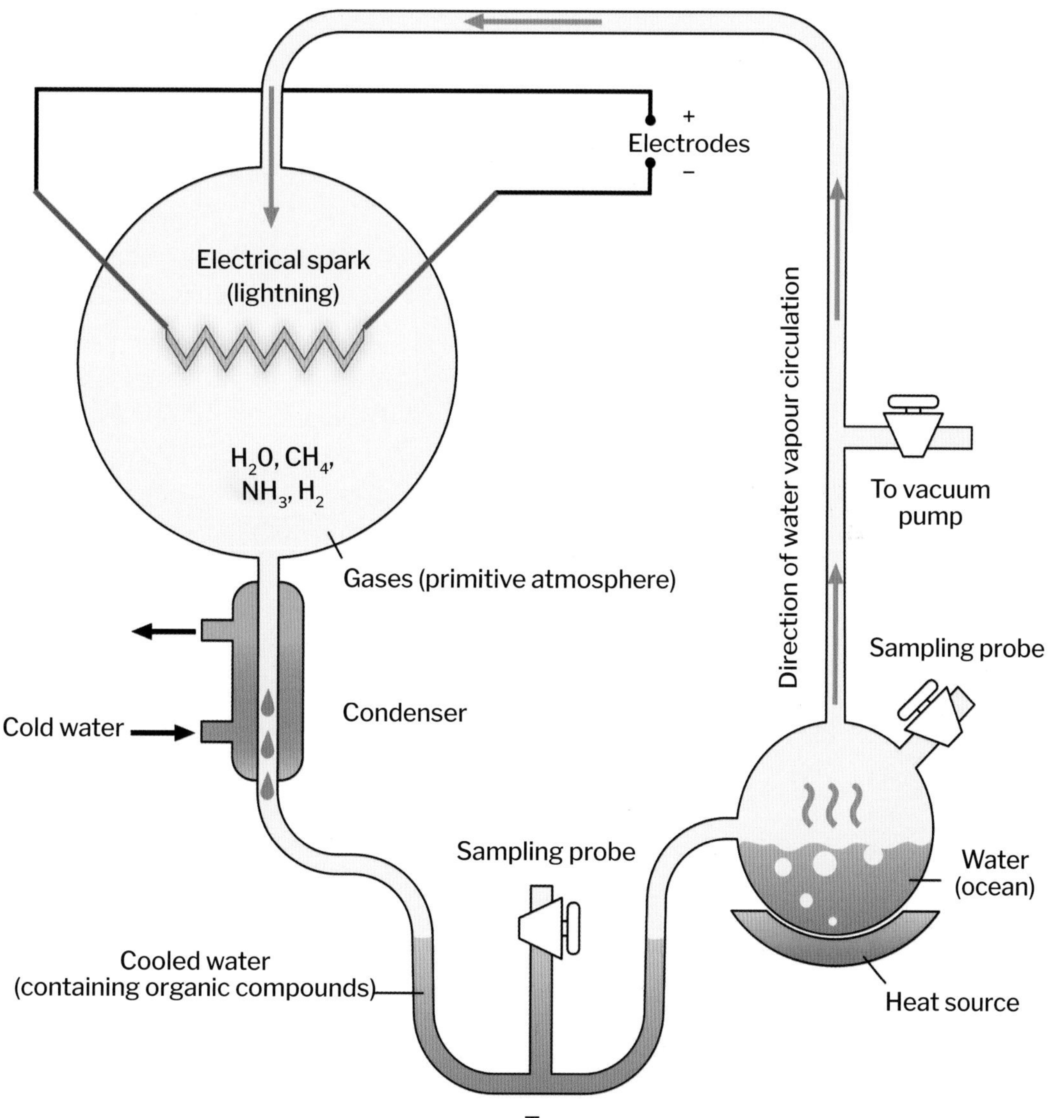

Chemistry set

All living things are made of organic chemicals – that is, chemicals that contain carbon in combination with hydrogen and often other elements, commonly including oxygen, nitrogen and phosphorus. Many of these organic compounds have very large molecules. One of the most important types is proteins. These are large complex molecules made from smaller units called amino acids. There are more than 500 amino acids, though only 21 are involved in genetics.

inside reproducing themselves. When a capsule becomes too large to be stable, it splits in two.

CHICKENS AND EGGS

All life now uses DNA to store genetic information and handle its reproduction. DNA stores a code that tells cells how to make proteins by arranging amino acids. But DNA can't form without proteins. It's hard to see how this sequence can start. One possibility is that life started with a related chemical, such as RNA, which can act as an enzyme and can also store information like DNA can. It could perform the function of both DNA and proteins itself. Some scientists think an 'RNA world' existed before DNA-based life. And maybe another, simpler, chemical kick-started life before we got to the RNA world.

OUT OF THIS WORLD?

Another possibility is that life didn't start off here but was delivered to Earth from space on meteorites. Amino acids have been found on meteorites from asteroids and a comet. If microbes of some sort, or the chemicals crucial to life, came from space that doesn't really answer the question of how life started – it just moves it elsewhere.

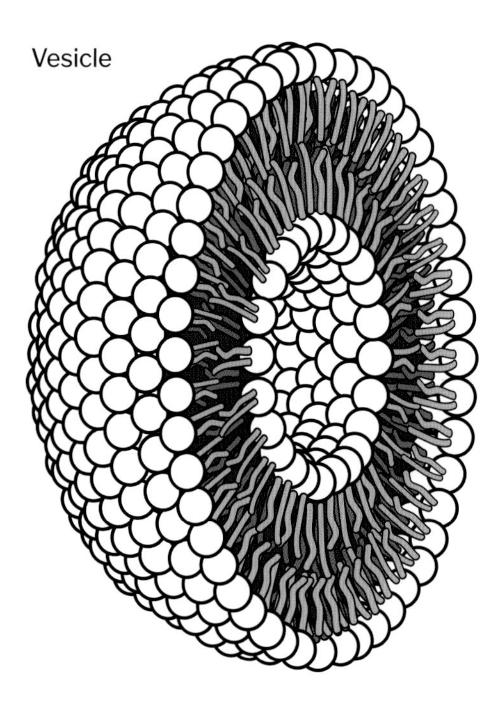

Cross-section of a capsule created by two layers of molecules hiding hydrophobic ends within the wall.'

21

Pushing together, pulling apart

Nuclear fusion and fission

The alchemists of the Middle Ages dreamed of changing one elemental metal into another. For a long time, it was deemed impossible. Similarly, the 'uncuttable' nature of atoms – their indivisibility – was accepted without question. But all that was undone by the discovery that radioactive decay does indeed involve changing one element into another, and splitting up atoms.

It turns out not only that atoms can be split to make different elements, but that it happens all the time. Likewise, they can be stuck together to make different elements – and stars make their living doing just this. Nuclear fission (splitting atoms) and fusion (joining atoms) are processes that produce enormous amounts of energy – star-making, world-destroying amounts of energy.

NEUTRONS ARE NEW

James Chadwick discovered the neutron, the uncharged particles in the nucleus of an atom, in 1932. The next year, Enrico Fermi and Leo Szilard proposed that it would be possible to produce a nuclear chain reaction, blasting atoms apart with neutrons, and using the extra neutrons freed to blast more atoms apart. This could produce energy for weapons or a nuclear reactor to generate power.

SPLITTING THE ATOM

Nuclear fission was first successfully achieved in Berlin in 1938 by Otto Hahn, Lise Meitner and Fritz Strassman. They demonstrated that bombarding uranium with a stream of neutrons broke the uranium atoms apart, making lighter atoms such as krypton and barium, while also releasing more neutrons and large amounts of energy. This was just before the start of World War II, and the three scientists were adamant that their work should not be used to build weapons. As German and Austrian Jewish scientists fled Germany, they took the secret of fission with them. Eventually, it ended up in the USA, where it was weaponized, partly because there was considerable fear that Nazi Germany was already working on nuclear weapons. Fermi and his colleagues produced the first controlled chain reaction in 1942. Soon after, the Manhattan Project took on the

Otto Hahn and Lise Meitner.

task of putting a nuclear chain reaction into a weapon, and it was downhill all the way to Hiroshima and Nagasaki.

Now, nuclear chain reactions are kept in check (usually) in nuclear power stations and used to generate massive amounts of energy to heat homes and power factories. Smaller nuclear batteries are used in submarines and in rovers sent to other planets. Only when things go wrong (such as in the disasters at Chernobyl in the USSR, and Long Island in the USA) are we reminded exactly how much energy is involved and how dangerous the materials can be.

PUTTING ATOMS TOGETHER

The process of nuclear fusion powers the stars, including the Sun. Under the intense gravity at the heart of a star, atoms are crushed together under such force and heat that they fuse. Initially, this is hydrogen fusing to form helium. The mass of four hydrogen atoms is slightly larger than the mass of the single helium atom they form. The extra mass becomes the energy that is released. In the Sun, the photons released by fusion slowly make their way to the surface and stream into space as the radiation that forms sunlight and its heat.

The core of the Sun has extreme conditions: temperatures and pressure beyond anything that can be easily replicated on Earth. Attempts to produce nuclear fusion soon targeted the creation of a hydrogen bomb: a bomb that would gain its massive power from fusing hydrogen. There have since been attempts to use fusion as the basis of a power supply but it's not been very productive so far. The first significantly productive reaction was produced in 1991, but to be really useful 'cold fusion' is needed – a fusion reaction that will take place at lower temperatures and

Nuclear fission chain reaction

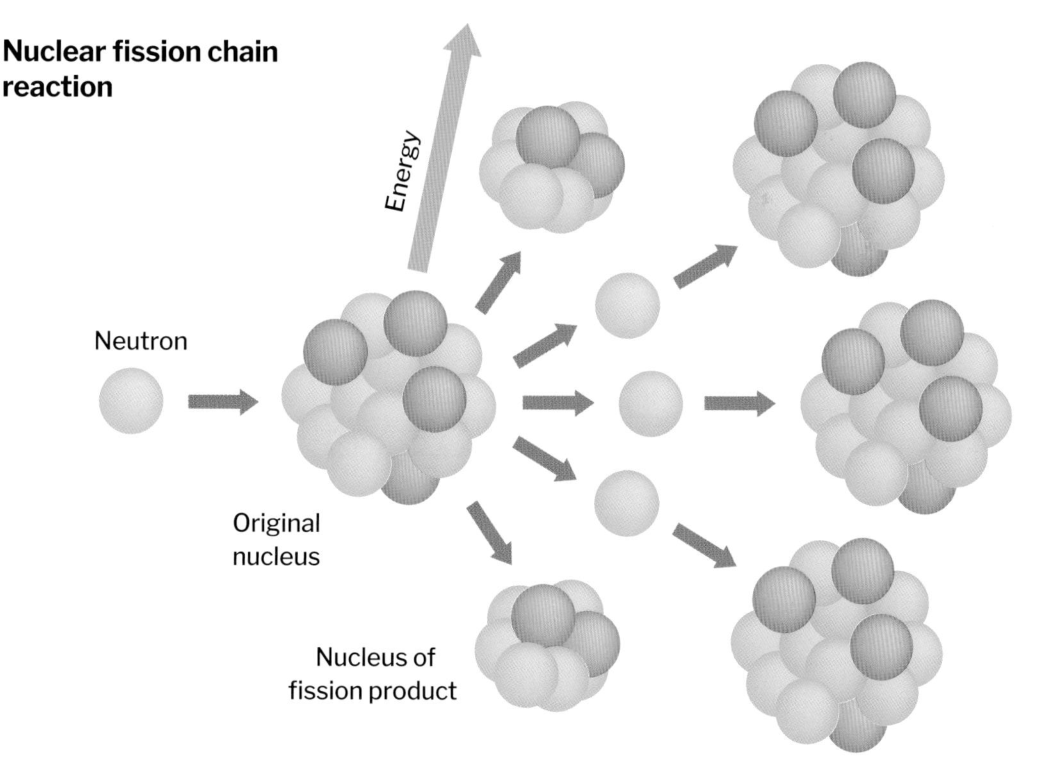

Neutrons freed by the first fission event impact other nuclei to start further fission events.

Where the energy comes from

Einstein's equation $E=mc^2$, shows that energy (E) and matter (in the form of mass, m) are ultimately equivalent. The amount of energy that can be released by destroying matter is its mass times the speed of light squared (c^2) – a huge number. As atoms are split apart, the mass of the products is slightly less than the mass of the original uranium atoms. The difference is released as energy. Even though the difference in mass is tiny, once it's multiplied by the speed of light squared, it becomes quite a large quantity of energy.

pressures. Fusion would be preferable to fission as a power source as not only does it produce far more energy than fission, it also produces less hazardous waste.

FUSION MAKES THE UNIVERSE

Human attempts at fusion would start with hydrogen and its variants, deuterium and tritium. Stars similarly begin by fusing hydrogen. When they have used up most of the hydrogen, though, there are too few collisions between hydrogen atoms to keep the process going. Then the helium they have created begins to fuse. The process goes on, producing progressively heavier elements such as carbon and oxygen. Stars can make elements up to iron. After that, fusion would take more energy than it releases, so it doesn't happen in the heart of a star.

A large enough star, when it has run out of elements it can fuse, can explode in a spectacular supernova event. While a star is working properly, fusion produces enough pressure from emitted photons to counteract the force of gravity pulling inwards on all parts of the star. When fusion stops, the pressure counteracting gravity is gone, and overwhelming gravity drags all parts of the star inwards. But there's nowhere for it to go, and the result is a massive explosion that tears the star apart as all the matter bounces outwards again. The explosion creates far greater pressure and heat than the functioning core of a star. In this moment, iron can be fused and heavier elements are produced. These, along with all the lifelong products of the star, are hurled out into space. They provide the material for making planets that will orbit future stars. Apart from the hydrogen atoms, every atom in your body was fused in a star or a supernova.

22

Starlight star bright

How stars work

Imagine yourself living in an age before the invention of the telescope. When you look at the night sky, what do you see? The Moon, obviously, which is large, often bright, and apparently changes shape in a cycle that takes about 29 days. Also, a lot of spots of bright white light – more than you're likely to see in the present day, as there was no light pollution back then. And probably a broad, milky band of light, the Milky Way, which doesn't resolve into separate stars.

Without modern knowledge of astronomy, you would have no idea how far away or large these points of light are, nor have any reason to suppose that they are not all the same distance from you.

FROM GODS TO GAS

Many early models of the heavens had the stars all hung on the surface of a hemispheric shell, like an upturned bowl over Earth. This suggests the stars are the same distance away and roughly the same size, but doesn't say what they are or how they work. Religion offered a simple answer: stars are lights hung in the sky by god(s), or are themselves gods. The Greek philosopher Thales of Miletus (624–546 BC) is the first person known to suggest that the Sun and stars are not gods but normal physical objects. The Sun, instead of being Apollo driving a chariot through the skies, was in his view a fiery ball hanging in space.

Anaxagoras (500–428 BC) went further, suggesting what the fiery ball might be. He was prosecuted in the 5th

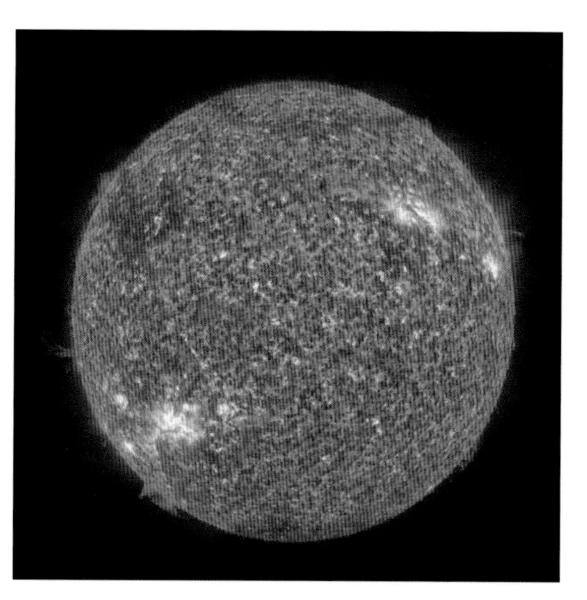

The Sun.

century BC for promoting his idea that the Moon is a chunk of rock and the Sun a burning rock rather than them both being gods. He also suggested that the Sun is the same as the stars, but we are close enough for Earth to be flooded by its light and heat while the other stars are too far away for this.

Little progress was made for more than 2,000 years. Then in 1609, Galileo looked through his telescope and discovered that the Moon has an uneven surface with features like those on Earth, such as rocky mountains; that the planets are disks and so could be other worlds something like ours; and that the Milky Way is a band of countless stars. That last discovery suggests that some stars are very far away. If we assume (it's not true, but a fair assumption for the time) that all stars are roughly the same size, if some are too dim and small to see as spots of light, that suggests they are further away than larger, brighter stars. Space acquired depth at the same time as it acquired many more stars.

> *'The composition of our own star and world is the same as that of as many other stars and worlds as we can see.'*
>
> Giordano Bruno, 1584

The telescope gave no clue as to what the stars are made of, though. Giordano Bruno, another insightful thinker persecuted for his work, suggested in 1584 that the Sun is a star like others

Fraunhofer lines in the spectrum of the Sun. Fraunhofer lettered the dark bands, and these have later been linked with different elements.

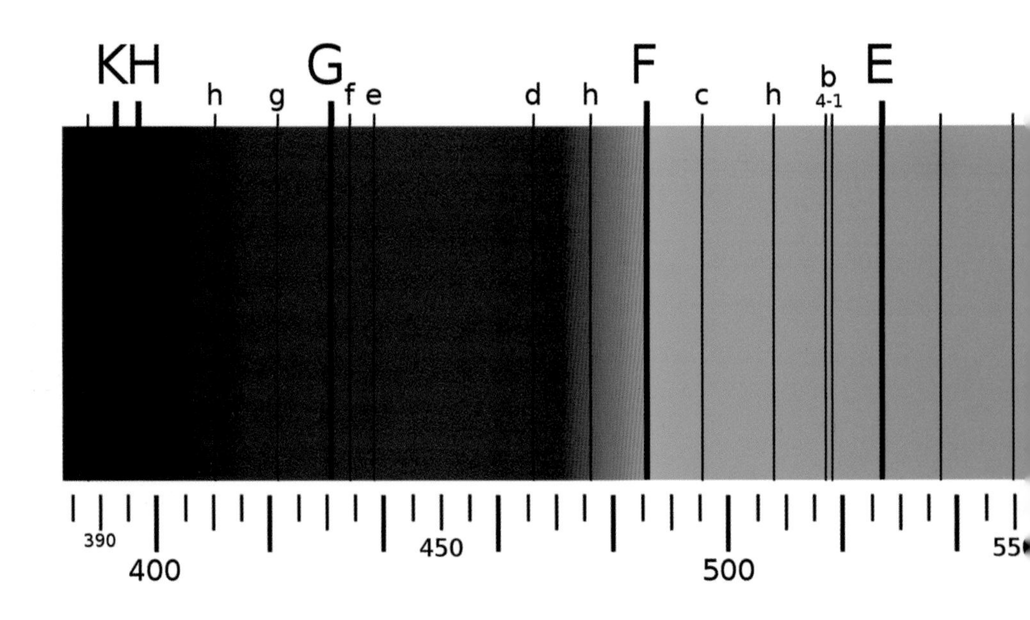

and that planets like Earth might orbit other stars. The notion that the Sun is an ordinary star grew more acceptable but was not confirmed until the 19th century and the invention of spectroscopy. Spectroscopy is a method of identifying elements by the wavelength of light that they absorb or emit. The link between elements and the spectra they absorb was pointed out in 1857 by Robert Bunsen and Gustav Kirchhoff. By examining the dark lines in the spectrum of sunlight, they were able to establish the elements present in the outer atmosphere of the Sun. Italian astronomer Angelo Secchi (1818–78) compared the spectra of around 4,000 stars and discovered that stars fall into a small number of distinct types. This definitively placed the Sun among the stars.

Spectroscopy tells us which elements are blocking light of some wavelengths as it leaves the Sun, but tells us nothing about the inside of the Sun. The first person to discover that the Sun is made mostly of hydrogen (73 per cent) and helium (21 per cent) was Cecilia Payne-Gaposchkin in 1925. An important question remained: how could a vast ball of gases produce the heat and light that power a solar system?

ENERGY FROM WITHIN

The answer came in 1929, from astronomer George Gamow. He proposed that within the core of a star, hydrogen atoms are forced together under immense pressure and heat, fusing through stages to form helium. This process, nuclear fusion, releases energy in the form of photons which slowly make their way to the surface of the star before pouring out into space. 'Slowly' can be very slowly indeed: a photon can take 100,000 years

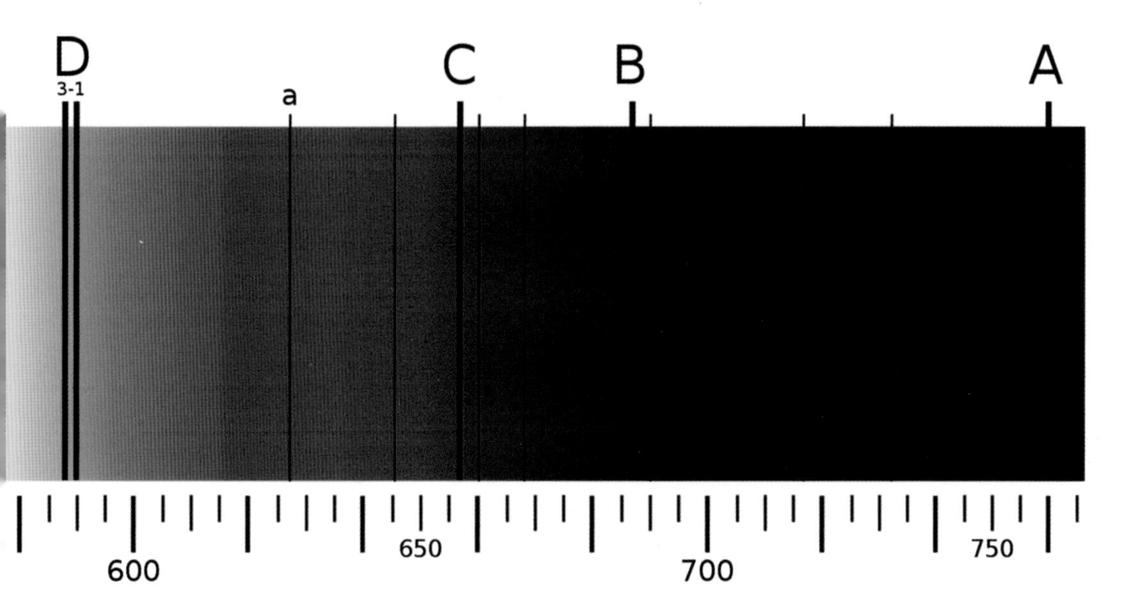

to reach the surface of the Sun before it can escape, streaming with others out into space. Photons are not trying to get out, and are not heading in any particular direction, so it is only when their random buffetings by other particles push them to the edge that they become part of the light and heat of sunlight. After all these years, it takes around eight minutes for a photon from the Sun to reach Earth. Only this final part of the journey is completed at the speed of light.

Constellation confusion

Cultures around the world have picked out constellations, making patterns and pictures in the shining dots of the stars. Seeing constellations encourages the idea that all stars are the same distance away (spots on an upturned bowl of sky). In fact, the stars in a constellation such as Orion are massively different distances from Earth, ranging between 500 and 2,000 light years away. (One of the 'stars' in Orion is actually a nebula, and is a vast dust cloud where stars are forming.) As soon as we put stars into a three-dimensional model of the sky, the constellations disappear entirely.

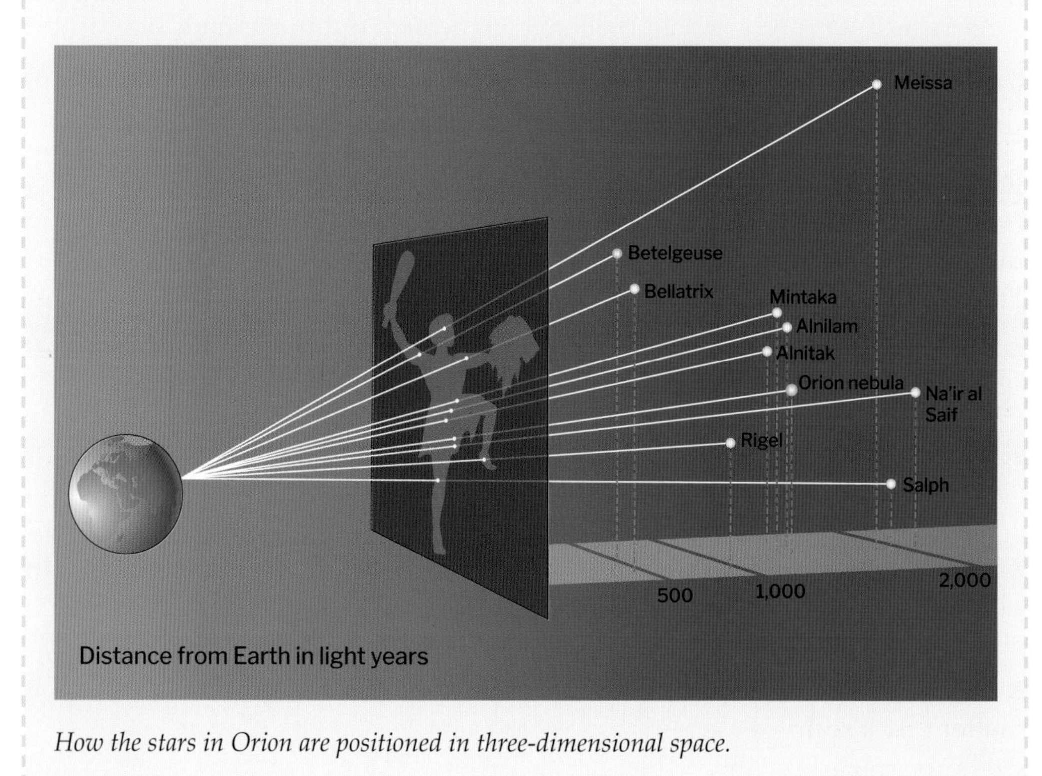

How the stars in Orion are positioned in three-dimensional space.

23

Dark forces at work

Dark matter and dark energy

When you look at the night sky, most of what you see is darkness – the dark emptiness of space between the stars. But just how empty is the darkness? It turns out the stars and other objects we can see make up only a tiny part of the universe. But what the rest is remains unclear.

HOW DO WE KNOW?

It might seem odd that astronomers can firmly state that there's a whole lot of invisible matter in the universe that we can't identify. We don't know what it is or where it is, but we're certain it's there. How?

There are billions of galaxies in the universe, which form clusters. These revolve, their speed depending on the distribution of mass in the cluster. From studying the location and mass of galaxies, it should be possible to predict the speed of the cluster's rotation. In the 1930s, Swiss astronomer Fritz Zwicky studied the rotation of the Coma galaxy cluster, hundreds of millions of light years away. Zwicky found the cluster to be revolving too fast for the mass of the matter that seemed to be there, judging by the light emitted. He introduced the term 'dark matter' for the invisible matter that it seems must be there.

Four decades later, American astronomer Vera Rubin showed that it's not just galaxy clusters but also individual galaxies that rotate more quickly than they should for the mass they appear to have. At the speeds galaxies and clusters rotate, they should be torn apart and scattered if they don't have considerably more mass than we can see. It appears from astronomers' calculations there is five or six times more matter present in the clusters than we know about. So while around a fifth of the matter in galaxy clusters is in the form of stars and other known objects, 80 per cent is in some other form.

LOOKING FOR LOST MATTER

If space is filled with matter that's invisible to us, the obvious question to ask is 'what is it?' There are several possibilities. It could be ordinary matter that we can't see because it doesn't emit visible light. In that case, it could be brown dwarfs, white dwarfs, neutron stars and black holes. Brown dwarfs are failed stars that didn't grow large enough to begin fusing hydrogen into helium and so emit light and heat. White dwarfs are the cooling cores of small- to medium-sized stars that have finished their life of nuclear fusion and no longer emit light and heat. Neutron stars are the super-dense remains of large stars that have imploded in a supernova. Only slightly less dense than black holes, they, too, emit no light. Black holes are the densest known areas in the universe that we 'see' as areas of emptiness and strong gravitational fields. The problem with this explanation is that there are almost certainly not enough brown stars, white stars, neutron stars and black holes to account for all the missing mass – remember that it must total four times the mass of objects we can see.

Another possibility is that dark matter is made of particles we haven't yet discovered. There are some particles that are predicted by theoretical physics but haven't been found yet. Some of these particles could be created in the Large Hadron Collider in Switzerland, but they would not be detected by the Collider. Their fleeting presence and then loss could be calculated from missing matter and energy in the system. Could there be enough of those to make up the missing mass? One theory suggests a 'hidden valley' – a parallel universe made up entirely of dark matter and that has little or no interaction with the universe we can see.

SEEKING THE INVISIBLE

One way in which astronomers hope to spot dark matter is by detecting flashes of gamma-ray radiation. This is very high-energy radiation, with a much shorter wavelength than visible light. It's possible that collisions between dark matter particles emit flashes of gamma rays. These could be detected by the Fermi Gamma-Ray Space Telescope, launched in 2008.

Another way of working out where dark matter lies is to use a technique called 'gravitational lensing'. This uses the fact that light is bent by gravity, as predicted by Einstein and demonstrated in 1919 (see page 40). A very massive object, such as a star or a black hole – or a concentration of dark matter – will bend light rays so that objects appear to be offset from their real position.

NOT ONLY MATTER BUT ENERGY

Dark matter isn't the only thing that astronomers are looking for – there is an even larger deficit. Around 67 per cent of the universe is now thought to consist of dark energy. (Einstein's theory of general relativity, page 186, shows us how energy and mass are interchangeable.) Adding this into the equation means that ordinary matter (not including dark matter) really accounts for only about 5 per cent of the universe. Most of the universe remains entirely and darkly mysterious.

The first evidence for dark energy came in 1998 with data from the Hubble Space Telescope. The telescope had been

The cloudy patches in this image of the galaxy cluster Abell 1689 are areas of dense dark matter, mapped from gravitational lensing images (but not visible under normal circumstances). The bright objects are galaxies, each with billions of stars. The Abell cluster is more than two billion light years from Earth.

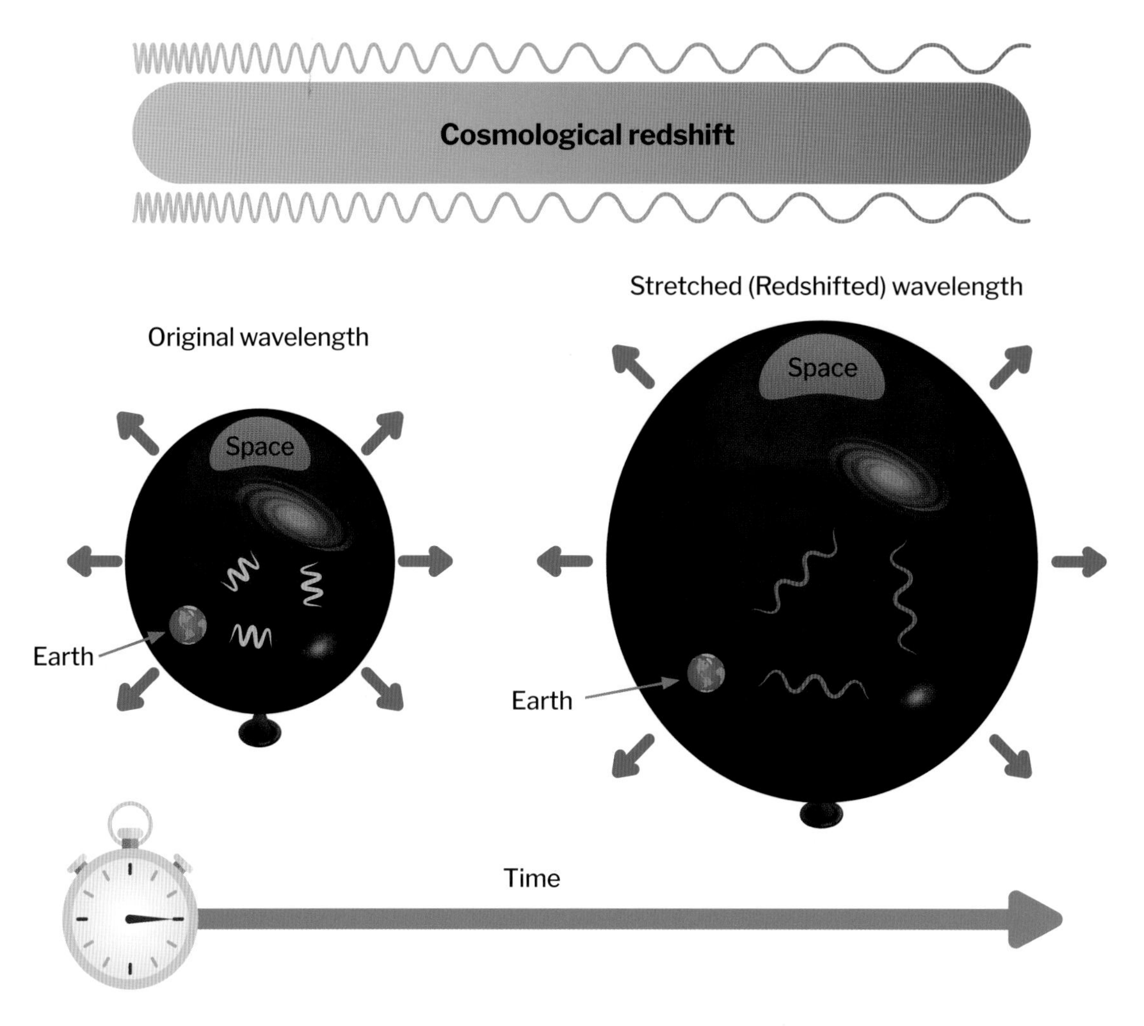

As time passes, space between Earth and a distant galaxy expands, dark energy forcing the locations further and further apart. Expanding space also expands the wavelength of light travelling between the two objects, producing a redshift.

searching for supernova remnants (the remains of exploded stars), and among those it found were six very ancient, distant supernovas. One of these first triggered the recognition of dark energy. Supernova 1997ff is ten billion light years away. Data from it revealed that the expansion of the universe began to speed up about seven billion years ago. The only way the expansion can accelerate is if some force is actively pushing the matter apart. That force is labelled 'dark energy' and it acts in the opposite way to gravity. The likely explanation is that around seven billion years ago, matter had become so spread out that gravity was no longer holding it all together firmly and dark energy got the upper hand, forcing it apart ever more quickly.

Dark energy seems to be evenly distributed, pushing galaxies apart at an even rate but having no local effects on gravity within galaxies and clusters.

A STICKY END?

The interaction of gravity and dark energy leads astronomers to consider three possible ways that the universe could end. One possibility is that dark energy will reach a limit and stop the acceleration

Redshift, blueshift

Astronomers can tell whether objects are moving towards us or away from us by examining light from them. Light from objects moving away is redshifted, which means it moves towards the red end of the spectrum and into infrared, with longer wavelengths. This is because the waves of light that have already left the object are stretched as the space between us and the object increases. Light from objects approaching us is blueshifted, the waves being pushed closer together producing a shorter wavelength.

Supernovas always emit the same amount of light. That makes it fairly easy for astronomers to work out from their apparent brightness how far away they are. By looking at the redshift of their light, it's possible to work out the rate of expansion of the universe over that time. And so by comparing distant supernovas, astronomers can build up a record of how the rate of expansion has changed over time.

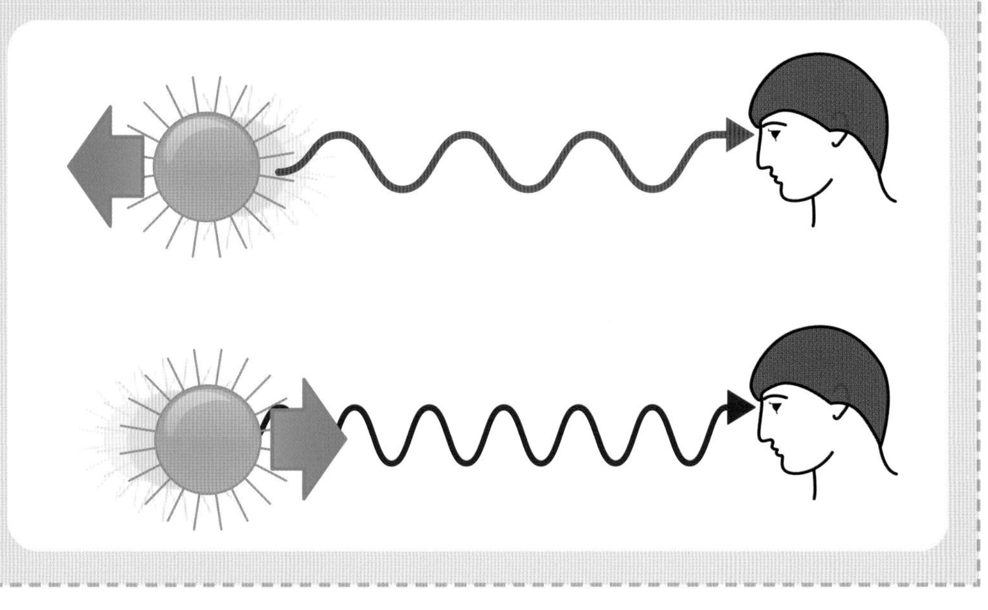

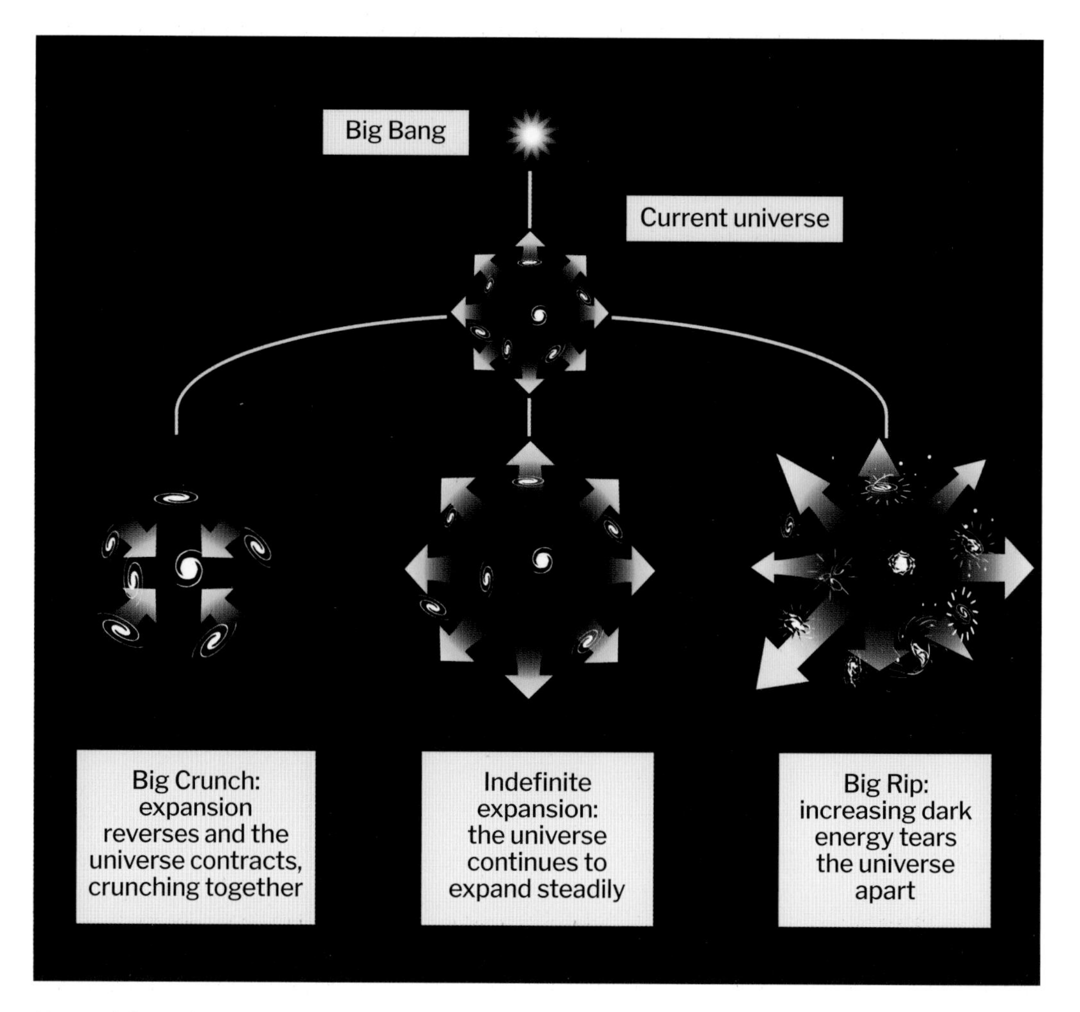

Fates of the universe.

of the universe. At this point, gravity could overcome the force of dark energy and draw all matter back in together, eventually producing a 'Big Crunch' – which could possibly then bounce into another Big Bang. Another possibility is that expansion will continue forever, but going more slowly. Eventually, all matter will disintegrate into energy. This scenario is called the 'heat death' of the universe. Finally, dark energy might win out over gravity, causing ever faster expansion until eventually galaxies, then stars and finally even atoms are wrenched apart. This is sometimes called the 'Big Rip'. But it's not an imminent worry. None of these is likely to happen for another 30 billion years or so.

24

From crust to core

The structure of Earth

The surface of Earth, on which we live, is just a thin crust covering thousands of kilometres of scalding rock and metal. We can't dig down beyond the crust, but have learned a lot about the interior structure of our planet.

A HABITABLE OUTSIDE

Earth's crust is a wafer-thin layer of hard rock, liquid ocean and gassy atmosphere, yet it's all we truly know. It's only about 10 km (6 miles) thick beneath the seabed and an average of 35 km (22 miles) beneath the land. It's thicker beneath mountains, at up to 50 km (31 miles) and thinnest in parts of the deep ocean. In all, it makes up less than 1 per cent of the volume of the planet. This is the fragile realm we inhabit.

After Earth formed, the surface cooled in contact with cold space, while gases including water vapour poured out of volcanoes. The water condensed into clouds and fell as rain for millions of years, eventually creating the oceans. As rocks weathered, and later living things altered the surface and the atmosphere, the crust developed the rich variety of environments we find around us today.

WHAT LIES WITHIN

Beneath our feet, the mantle stretches far below the crust, finally meeting Earth's core. The mantle is made of rock and the core mostly of metal. The two separated early in Earth's history, with the heavy, molten metal seeping between rock particles, drawn by gravity to collect deep within. While the crust is active, varied and has changed considerably over Earth's history of 4.5 billion years, the inside of the planet has barely changed since it began to differentiate into core and mantle.

A FLOWING CLOAK OF ROCK

The mantle is thick, viscous rock, churned slowly by convection currents. Just below the crust, it's heated to 1,000°C (1,832°F), but near its boundary with the core, 2,880 km (1,798 miles) below, the temperature rises to 3,700°C (6,692°F). It is this scalding rock, called magma, which wells up at rifts in Earth's surface and spews from volcanoes.

The mantle makes up more than 80 per cent of Earth's volume. Most of it is

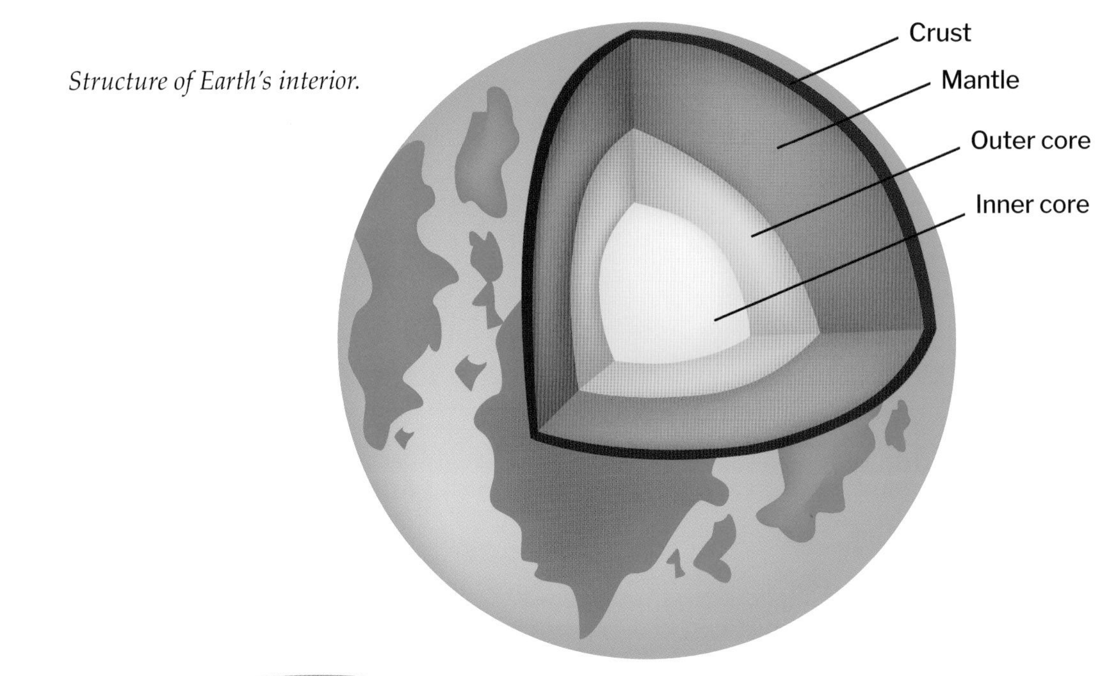

Structure of Earth's interior.

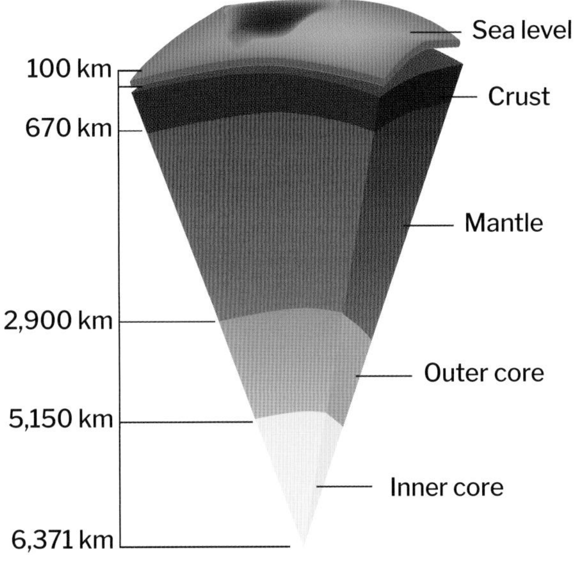

solid rock, despite being so hot. At some places it's more liquid, and flows slowly. As it creeps along, it drags the slabs of Earth's crust with it, so that the surface is constantly edging its way around the globe (see page 111). This movement is so slow that it takes millions of years to make any noticeable difference to the arrangement of land and sea.

The mantle is divided into layers itself, with the upper and lower mantle separated by a transition zone. The very top of the upper mantle, with the crust, forms the lithosphere, a layer of hard, solid rock. A much thicker layer of hot, semi-mobile rock below it is the asthenosphere. Within the upper mantle, convection currents allow heat and rock to circulate slowly. Rock heated low down by the planet's core gradually rises and cooler rock falls to take its place. The rock is made mostly of silicates — rock containing silicon and oxygen, including green olivine, red garnet, and black pyroxene. A transition zone separates the upper mantle from the lower mantle, which is made of much thicker, more solid rock. The lower mantle possibly doesn't move at all — or it might have its own convection currents, or might exchange with the upper mantle. No one really knows. Although it's hotter, the intense pressure so far down keeps the rock solid, with no room for the molecules to move around.

RIGHT TO THE CORE

At first, Earth was a ball of hot rock, roughly the same throughout. Slowly warmed by a mix of left-over accretion heat and radioactivity deep within, it eventually reached the temperature at which iron melts. Then Earth began to differentiate. The heavy, molten iron seeped down to the centre and the lighter gases and water escaped towards the surface, leaving rock in between.

Earth's core is made mostly of iron and nickel and again it layers. Around 3,485 km (2,165 miles) thick altogether, it's divided into the outer and inner core. Temperatures deep inside Earth fluctuate. They are hard to estimate and impossible to measure, but seem to range between 4,400°C (7,952°F) and about 6,000°C (10,832°F). The outer core is a layer 2,200 km (1,367 miles) thick, ranging in temperature from 4,500–5,500°C (8,132–9,932°F). It has low viscosity, so it's quite fluid, churned with convection currents. The flowing metal of the outer core produces Earth's magnetic field (see page 120).

The hottest area of Earth is the boundary between the outer and inner core, called the Bullen discontinuity. Here, the temperature reaches a scalding 6,000°C (10,832°F), as hot as the surface of the Sun. Below this boundary, the inner core is solid. Again, the pressure is too great for the iron and nickel atoms to move around as a liquid, being nearly 3.6 million times atmospheric pressure. It's slightly cooler, at 5,200°C (9,392°F). There's probably an *inner* inner core, also made of iron and nickel and still solid. The difference is that the atoms line up in a different direction (east–west rather than north–south) from those in the inner core. It might be only 500 million years old, caused by some unknown geological event, and remains largely mysterious.

LOOKING WITHIN

It's impossible to drill down even to the mantle. All our drills are restricted to the crust. Scientists find out about the deep insides of Earth by looking at materials

Not going to happen

The solid inner core grows by about 1 mm a year as Earth slowly cools. It grows by crystallizing metal from the outer core. It isn't a scenario to worry about. There isn't time for the whole of the outer core to crystallize before Earth ends through other ways: it would take around 91 billion years to fully harden, while the Sun will expire in around five billion years.

Inge Lehmann.

that have been brought to the surface, in volcanic eruptions, as inclusions in other rocks, and by measuring how seismic waves (the waves of energy produced by earthquakes) move through Earth. The Danish geologist Inge Lehmann discovered Earth's solid inner core in 1936, analysing the seismic waves produced by a large earthquake near New Zealand seven years earlier. How quickly waves propagate reveals a lot about the composition of the materials they're passing through and where the boundaries lie between them.

FANCIFUL INNER SPACES

Earlier ideas about the structure of Earth included that it was hollow (perhaps with other races of beings living there), that it contained Hell, or that a series of hidden funnels and tunnels threaded through the land under our feet. In 1664, German scholar Athanasius Kircher imagined a gaping hole at the North Pole sucks water in towards Earth's blazing central fires. There, he thought, it was heated and ready to be spewed out at the South Pole. Further tubes, channels and underground lakes contain magma, which sometimes comes out in volcanoes. Even the astronomer who explained comets, Edmund Halley, suggested that Earth is really hollow with three more spherical shells within the one that we inhabit. He didn't go as far as the American physician Cyrus Reed Teed, though, who claimed in 1869 that Earth is not only hollow but that we live on the inner surface.

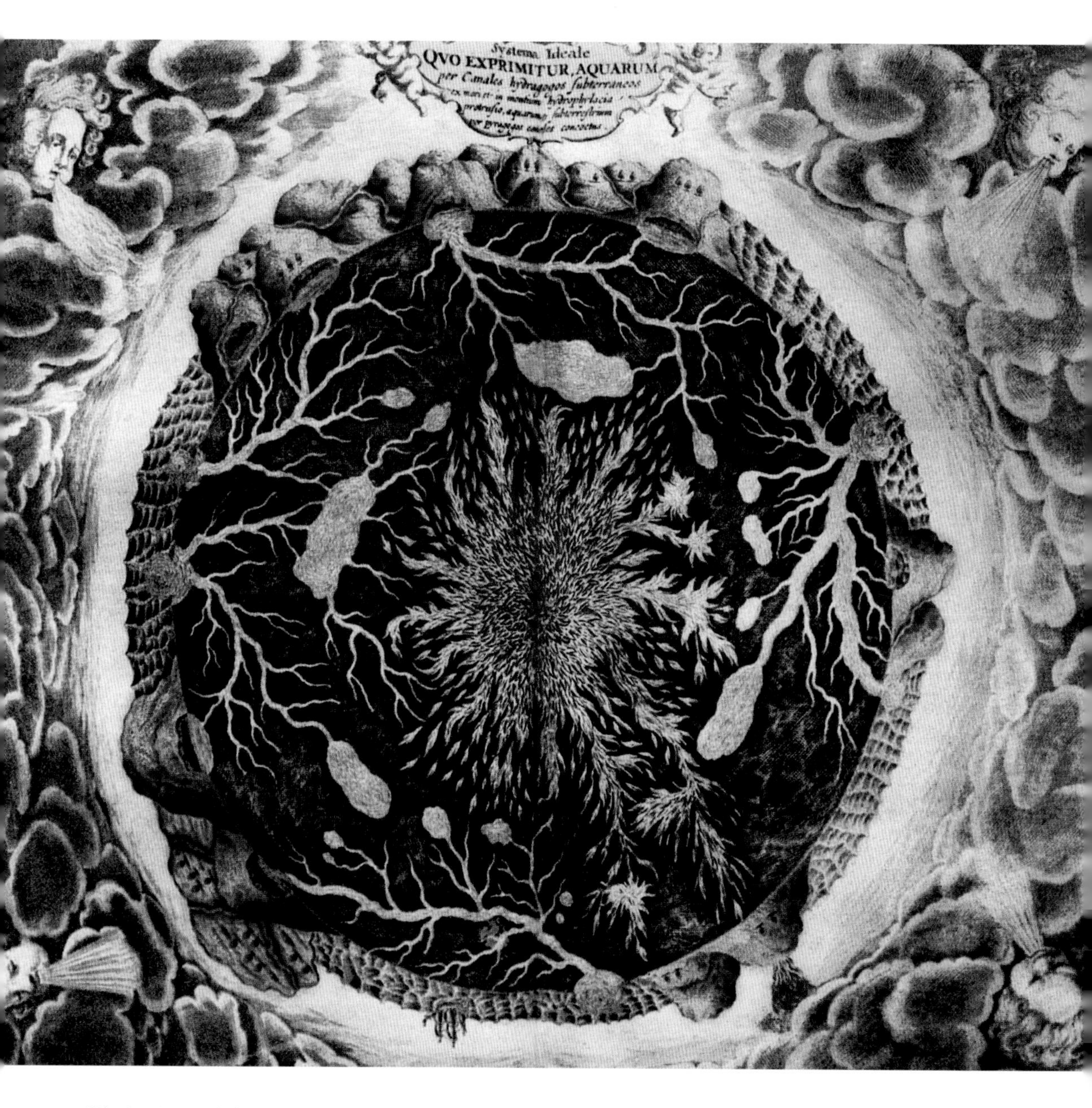

Kircher's Earth had internal channels for water and magma, and a central fire.

25

Smashing plates

Tectonics

Soon after Europeans discovered the existence of the Americas and mapped the eastern coasts, people noticed that South America would fit snugly against Africa if it weren't separated by the Atlantic Ocean. It took until the 20th century to discover that the two were, once, fitted together.

FITS AND FOSSILS

Alfred Lothar Wegener, a meteorologist and geophysicist, was fascinated by the fit between Africa and South America. He looked for and found other evidence that the lands were once joined. Ancient slabs of rock, called cratons, and belts of mountains are also now divided by the Atlantic Ocean. The cratons, two billion years old, have chunks left behind in South America while lying mostly in Africa. The Caledonian / Acadian mountains are neatly split between North America, Greenland, Great Britain and Scandinavia.

Even fossils suggest the lands were once joined, with the fossil remains of identical plants and animals found on lands now widely separated, including Africa, India and Antarctica. Some of these were large animals that could not possibly have swum across an ocean. The dates of these fossils, 300–240 million years old, tell us how recently the lands were joined together. Furthermore, the fossils suggest the areas where they were found once had a very different climate. Both the frozen Antarctic and northern Norway yield fossils of tropical plants which could certainly not survive there now.

Wegener suggested that at one time the continents had been joined together in a 'supercontinent', but over millions of years they separated and drifted to their current positions, a process he called 'continental drift'. The 'supercontinent', called Pangaea, incorporated most of the land on Earth. Unfortunately, Wegener could suggest no mechanism by which continental landmasses might have moved. When he published his work in 1915, it attracted little support and a fair amount of ridicule.

DRIFTING

Evidence for how continental drift happens finally emerged in the 1950s and

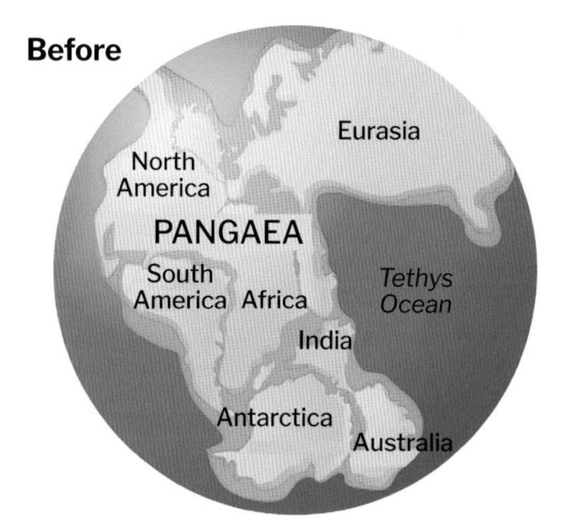

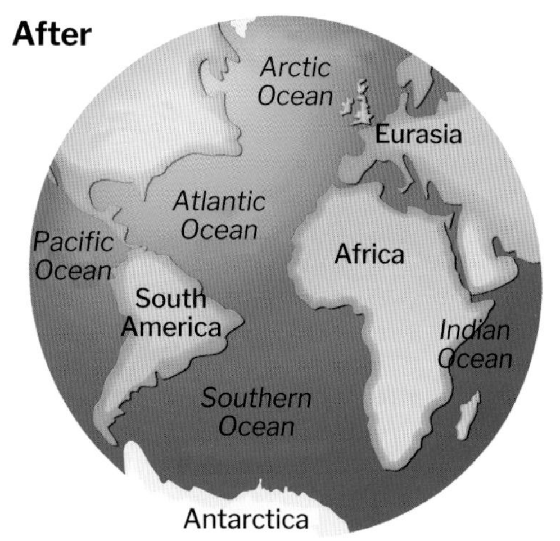

The current continents were joined in the supercontinent Pangaea, which began to break apart 200 million years ago.

1960s. With the ability to map the seabed with sonar and radar came the revelation that vast mountains, valleys and canyons lie hidden beneath the oceans. Mapping showed a rift along the middle of the Atlantic Ocean, fringed by mountains – a rift that suggests Earth's surface is slowly pulling apart. This allows magma to leak up from the mantle, hardening into new seafloor. The rift extends to Iceland, where some volcanoes on land also pour out new rock, stretching the island. Some volcanoes off the coast of Iceland form new islands in the same way.

BROKEN PLATES

There are other areas of Earth where chunks of the crust are pulling apart, too. It turns out the entire surface comprises slabs of crust, called tectonic plates, that sit on top of the mantle. There are 13 large plates and many smaller ones, all irregularly shaped and moving independently. Convection currents in the upper mantle slowly shift the tectonic plates, so that over millions of years slabs of land move around. Where new rock wells up in the ocean, it pushes aside the existing seabed. At the edges of the ocean, the oceanic crust is pulled under the lighter crust of the continental landmasses in a process called subduction. Vast slabs of seabed are dragged down, with some of their water, to be melted and remixed in the mantle.

Just inland of the subduction zones, volcanoes form, releasing lighter material that rises from the melting oceanic crust. Through these volcanoes, water, carbon dioxide and lava (magma) return to the surface.

AT THE EDGES OF THE WORLD

Subduction zones are common, active boundaries between tectonic plates, but they are not the only type of boundary. Geologists recognize three types of boundary between plates. A subduction zone is a convergent boundary, with

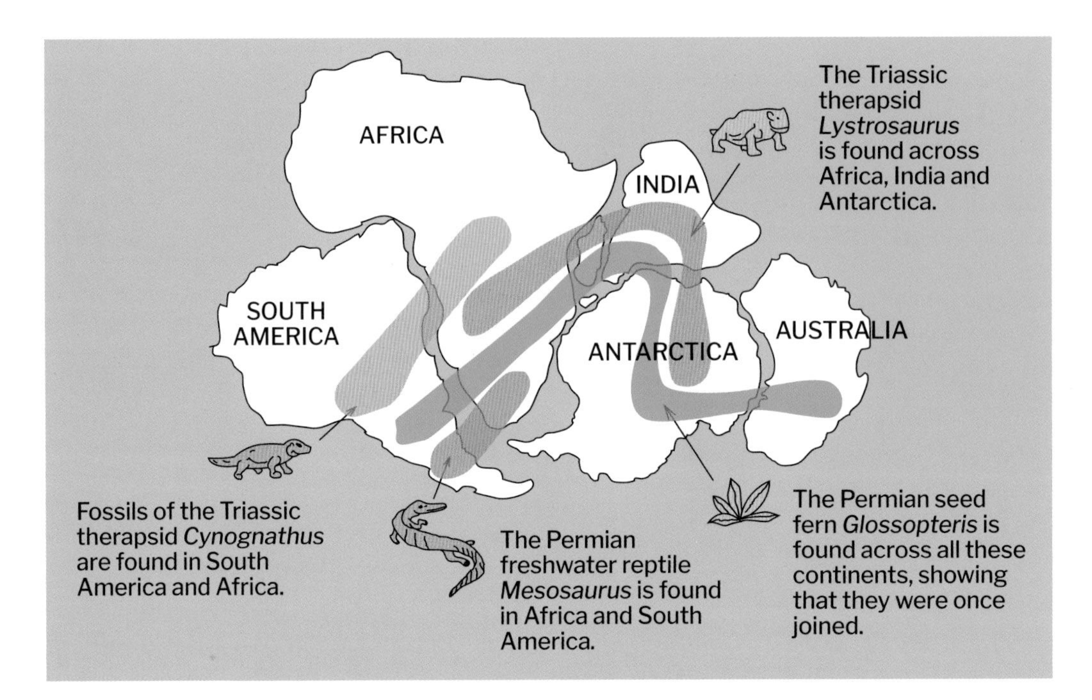

The distribution of fossils shows how landmasses were once joined.

two plates moving towards each other (converging). Another type of convergent boundary has two continental plates colliding. Typically, this pushes up mountains as the rock buckles and piles up. Convergent boundaries are often marked by a long string of volcanoes and frequent earthquakes. The 'ring of fire' is a border of volcanoes and earthquake zones that circles the vast Pacific plate. It runs all along the west coast of South and North America, through Japan, the Philippines, Indonesia and New Zealand.

At a divergent boundary, plates separate, and magma leaks up to form new rock. These are rifts, likethose on the seabed and in Iceland. At a transform plate boundary, plates move alongside each other in opposite directions. Sometimes they grate and snag and then later suddenly jolt as they move on again, causing earthquakes.

ON THE MOVE

The movement of tectonic plates explains volcanism, earthquakes and continental drift, and so forms a kind of unifying theory for geology. Over much longer timescales, it explains how continents form, break apart and reform, and how oceans grow between blocks of land. At several points in the past, most of Earth's land has been clumped into a supercontinent.

Individual landmasses have also moved around the globe. Finding fossils of tropical plants in Antarctica doesn't necessarily mean the region near the South Pole once had tropical conditions.

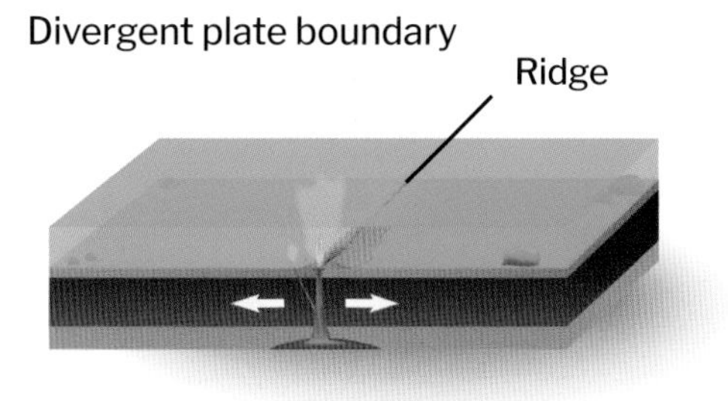

Transform plate boundary

Earthquakes

Convergent plate boundary

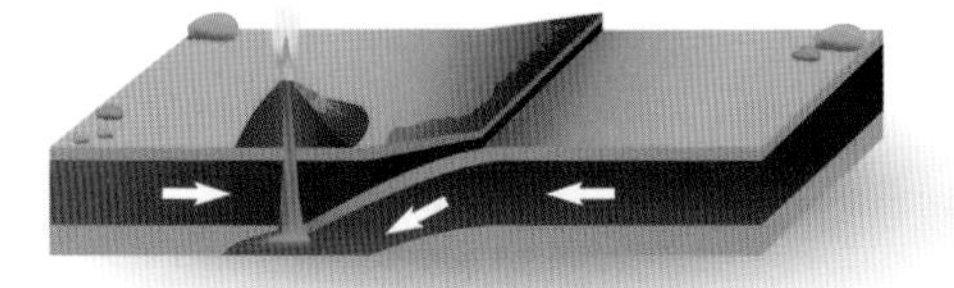

Types of plate boundaries.

In fact, the land now situated at the South Pole was once near the equator but has moved, though the poles have also been temperate and ice-free in the past. When fossils of marine organisms are found on mountains, that tells us that the rock now on a mountain was once the seabed, near a coast that has collided with another and been pushed upwards into mountains. The Himalayas formed when India – once an island – crashed into the Eurasian landmass over millions of years. The movement continues. The Atlantic Ocean grows wider by a few millimetres each year, and at some time in the distant future, the continents will have rearranged themselves once again into a supercontinent.

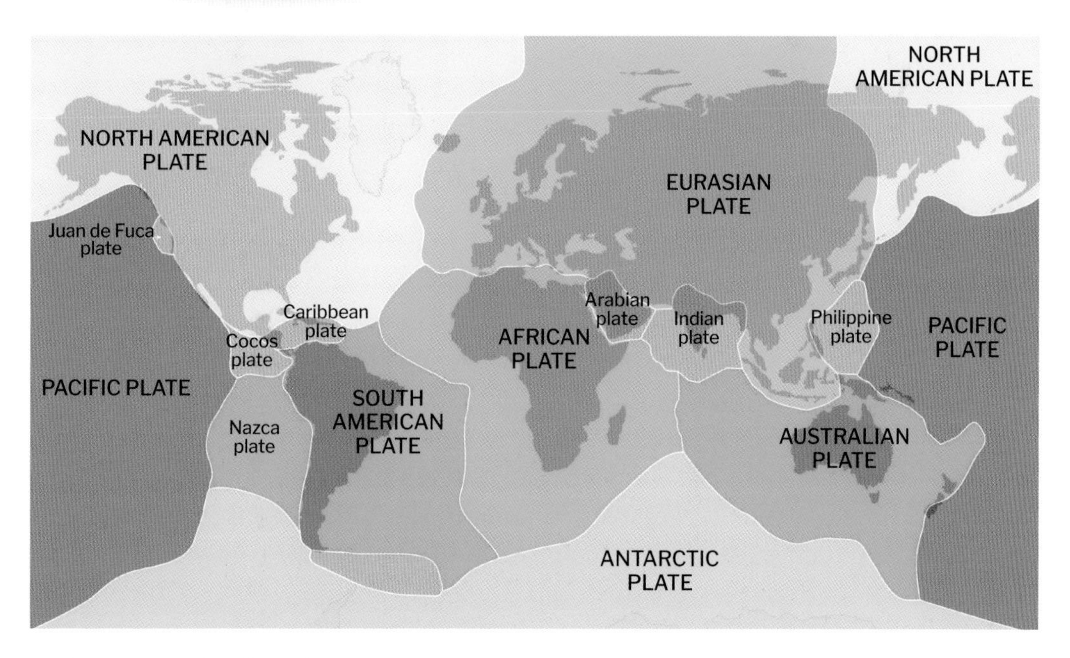

Major tectonic plate boundaries.

26

Heat and gases

The greenhouse effect

People most often think of the greenhouse effect in relation to the current climate crisis caused by human activity, but the natural greenhouse effect has created Earth's climate over billions of years. It's a mechanism by which a planet maintains a stable temperature and retains heat – but it can be catastrophic if it's too effective.

IN THE GREENHOUSE

In 1859, Irish physicist John Tyndall began studying the ability of gases to retain heat. He found that of the gases in Earth's atmosphere, water vapour and carbon dioxide retain a lot of heat, while nitrogen and oxygen retain virtually none. Without water in the atmosphere, he concluded, Earth would be perpetually frozen. In 1895, the Swedish chemist Svante Arrhenius calculated the effects of reducing and increasing concentrations of carbon dioxide in the atmosphere and declared that raised levels will produce higher temperatures and depressed levels will lower the temperature.

EARLY GREENHOUSE IN A COOL EARTH

Early in Earth's life, the Sun was much less bright, producing less energy than it does now. If Earth depended only on heat from sunlight, it should have been a frozen wasteland. Instead, life emerged and flourished. Earth could become hospitable because its atmosphere was rich in water vapour and carbon dioxide – both greenhouse gases. The gases trapped heat, preventing it escaping into space. Microbes added methane to the evolving atmosphere, too. This helped to warm the world, making it more habitable and kick-starting Earth as a planet crowded with life.

THE BLANKET BECOMES THREADBARE

But life was its own undoing. Photosynthesizing microbes called cyanobacteria evolved which used carbon dioxide and released oxygen. The cyanobacteria were very successful, and the production of oxygen increased rapidly, in a 'Great Oxygenation Event'. The loss of carbon dioxide thinned the protective blanket of greenhouse gases, cooling Earth. It cooled so much that

The greenhouse effect: a layer of gas traps heat near Earth, reflecting it back to the surface.

the entire planet was covered in a thick layer of ice, with perhaps just a few pockets of liquid water. Sometimes called 'Snowball Earth', this period of extreme cold lasted hundreds of millions of years, from around 2.4 to 2.1 billion years ago. Luckily, volcanic activity continued, slowly releasing enough carbon dioxide to re-establish the greenhouse effect. Earth warmed enough for the ice to melt, and life took off again.

A new balance kept Earth habitable, though there were further Snowball Earth events, again probably triggered by increased photosynthesis with the evolution of land-based plants.

SWINGS AND ROUNDABOUTS

The balance of carbon dioxide and oxygen in Earth's atmosphere has varied a lot over the planet's history – even over its relatively recent history since the evolution of life on land. This has affected the climate, with periods when Earth has been much hotter than it is now and periods when it has been much colder. We know that life can survive higher and lower temperatures than we have now. Organisms evolve to suit the prevailing conditions, or they die out. Those that lived in the hot, dry conditions of the Triassic, 240 million years ago, could not survive now, as we could not survive in the conditions they thrived in. Usually, the climate changes slowly, giving organisms time to move around and adapt. Most mass extinction events (see page 182) have been caused by rapid climate change.

TOO MUCH AND TOO LITTLE

Earth's greenhouse effect helps to even out the temperature. The planet Mercury, with no atmosphere, has extremes of temperature. The side facing

Snowball Earth.

away from the Sun can drop to –180°C (–290°F) while the side facing the Sun is a scorching 427°C (800°F). On the other hand, Venus – which is not as close to the Sun as Mercury – has an atmosphere made mostly of carbon dioxide and an extreme greenhouse effect. The planet is roughly 390°C (700°F) hotter than it would be with no greenhouse effect, at 475°C (900°F).

From trees to coal

During the Carboniferous period, 359–299 million years ago, great forests grew over much of the land. These locked in carbon, but when they died, instead of decaying and releasing their carbon, many were fossilized. They became the coal reserves we now burn. By the end of the Carboniferous, Earth had cooled almost to modern temperatures, but then the microbes that break down wood evolved and the carbon balance began to be restored, with carbon released back into the carbon cycle when trees died.

27

Too hot, too fast

The climate crisis

The evidence of a climate in crisis is all around us: the increasing frequency of extreme weather events, melting ice, rising sea levels and rising temperatures are all evidence that the climate is changing. While Earth's climate has always changed, it has rarely changed as quickly as it is changing now, tied to increasing levels of greenhouse gases in the atmosphere.

BURNING THE WORLD

Since the Industrial Revolution began the large-scale use of fossil fuels (coal, gas and oil), the average global temperature has risen to levels not seen in the last 100,000 years. This period spans the whole of human civilization: our civilization is built around organisms that flourish, and weather patterns that prevail, at lower temperatures.

The average global temperature on Earth has risen by just at least 1.1°C (1.9°F)

Rising global temperature, 1880–now.

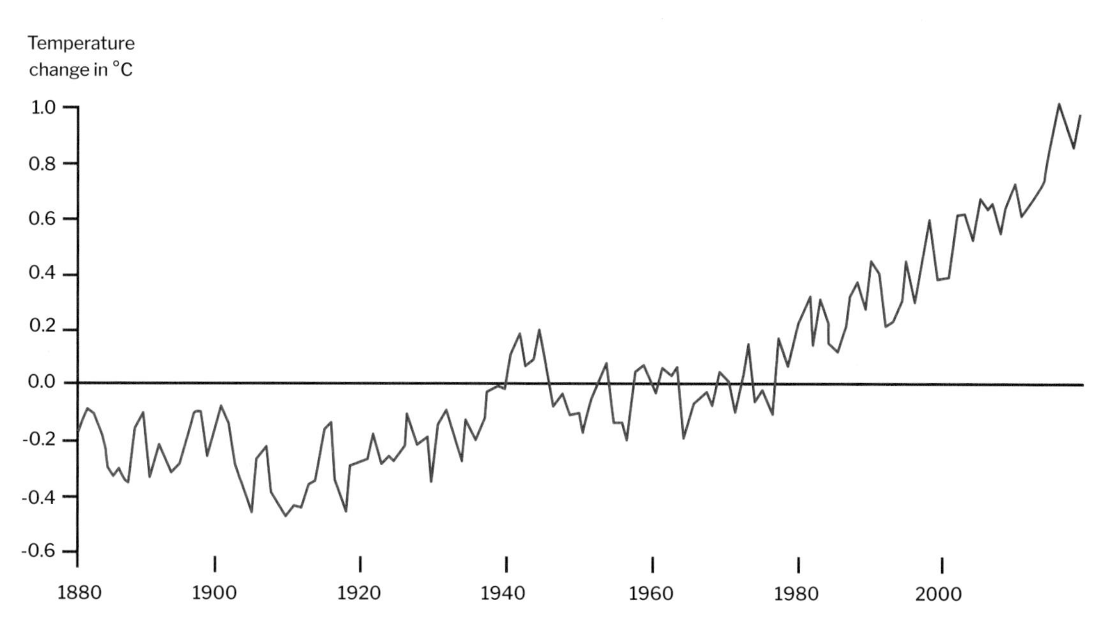

since 1880, the earliest date for which we have reliable worldwide temperature records. Two-thirds of the increase has happened since 1975, at a rate of roughly 0.15–0.20°C per decade. Using a baseline of 1951–80, global temperature has increased 1°C, and it's unlikely to remain below 1.5°C (2.7°F). The rise is the result of increased greenhouse gases, principally carbon dioxide and methane. We have caused this increase by burning fossil fuels, intensive farming and destroying forests that remove carbon dioxide from the atmosphere. Although 1.5°C doesn't sound much, it's enough to disrupt established weather patterns. It's not equally distributed: a steep rise in some areas can be offset by cooler temperatures elsewhere which brings the average down. The temperature in the Arctic has risen around 4°C (7.2°F), enough to melt a lot of ice, raising sea levels.

WHAT ARE WE FACING?

If the temperature rise isn't contained, humanity faces a dire situation. Failing to act is predicted to raise the average global temperature by 5°C (9°F), which would have catastrophic effects, making great swathes of land unusable for farming or living on. The disruption to weather patterns will be complex and have far-reaching impacts, too.

Melting polar ice and glaciers will raise sea levels, flooding low-lying areas and many coastal cities. The sea level has already risen around 15 cm (6 in.) in the last 100 years, and is likely to rise 1–2 metres (3–6 ft) by 2100. That's enough to flood an area the size of Alaska.

Raising the temperature of the sea changes global ocean currents and they in turn affect the weather. If a rise in temperature means the pattern of currents in the north Atlantic Ocean is disrupted, the Gulf Stream, which delivers warm water to Europe, is likely to collapse. This will bring colder winters to western Europe.

While there's more water in some places, there will be less in others. Drought and high temperatures will make some areas uninhabitable or unfit for growing food. The loss of glaciers inland will cause water shortages in areas which depend on meltwater for drinking, farming and industry.

WE ARE NOT ALONE

Humans share the planet with millions of other species. We rely on established ecosystems – the complex networks that connect organisms in an environment. Climate change threatens other species, too. Many can't now freely migrate to find suitable conditions as large stretches of land have been built on. Animals can't cross them unless we leave wildlife 'corridors' joining regions of wild habitat. Climate crisis threatens the insects we rely on to pollinate crops and the fish we pull from the sea; it might favour insects and microbes that destroy crops, removing their predators.

TIPPING POINTS

Climate scientists worry that there are tipping points beyond which climate change will accelerate unstoppably. If the sea heats to the point where all coral reefs die (about 3°C (5.4°F) rise), many reef-dwelling fish and crustaceans will die out, disrupting marine ecosystems. As permafrost melts in the far north, methane will be released into the atmosphere that will drive further heating. Melting ice doesn't only increase sea level, it accelerates heating. White ice reflects light and heat but dark sea absorbs it, so melting ice will also increase temperatures. Once these tipping points are reached, there is no stepping back, no reversing their impact. Some of the results of carbon dioxide we are releasing now won't be fully felt for hundreds of years – some will be felt within a decade, and some are being felt already.

Flooding in Germany in 2021.

28

Atoms lining up

Magnetism

A compass has a needle that points towards the north because it aligns with Earth's natural magnetic field. That the naturally magnetic 'lodestone', containing the mineral magnetite, attracts some types of metal was noticed in ancient China, India and Greece more than 2,000 years ago, and has been used for navigation for nearly 1,000 years. Magnetism is more important than just as a pre-industrial satnav, though.

MAKING MAGNETS

While lodestones are naturally magnetic, most of the magnets you will encounter today are deliberately magnetized metal. The most readily magnetized is iron. A piece of iron, as the Ancient Chinese discovered, can be magnetized by repeatedly stroking it in the same direction along a piece of lodestone. This is because a piece of iron or other magnetic metal has many tiny magnetic domains, like micromagnets, within it. In unmagnetized metal, these are higgledy-piggedly, with their north and south ends pointing in all directions. The result is that there is no overall direction of magnetic force. But a lodestone, or another magnet, can be used to turn the magnetic domains, eventually getting them all lined up to point the same way. All the little magnetic fields reinforce each other, producing an overall magnetic field for the whole object.

A magnet has two poles, called north and south. Like poles repel each other, and opposite poles

A Chinese lodestone spoon could swivel to align itself north–south.

attract each other. When two magnets are brought together, the north pole of one magnet will attract the south pole of the other. The north poles will repel each other. A compass works because Earth itself has a magnetic field, and acts as a giant magnet. The north poles of all the magnets in the world are attracted to the pole we call 'north' on Earth – but in fact, the north pole is actually the south pole of Earth's magnetic field!

MAGNETIC EARTH

Not much was known about magnetism until the English physician and scientist William Gilbert (1544–1603) published *De magnete*, in 1600. In it, he made the astonishing revelation that Earth itself is a giant magnet. We now know the source of this magnetism is movement of electrons in the molten iron core of the planet.

Earth's magnetic field is vast, spreading far out into space. It helps to deflect the solar wind, the stream of charged particles coming from the Sun. It acts as a shield, directing the particles around Earth. This prevents harmful radiation reaching the surface and helps keep Earth's atmosphere intact – it could otherwise be stripped away, leaving a nearly airless planet that would be hostile to most life. If our planet didn't have a molten iron core and a magnetic field, it might not be inhabitable at all.

Earth's magnetic field protects most

No fixed abode

Earth's magnetic north pole isn't in the same place as the geographic North Pole (the most northerly point of the globe). Instead, it's about 1,600km (1,000 miles) south of the North Pole, somewhere in Canada. The geographic North Pole is the end of Earth's rotational axis. The magnetic north pole doesn't even stay in the same place: over the last 150 years, it's moved around 1,000 km (620 miles) towards Siberia.

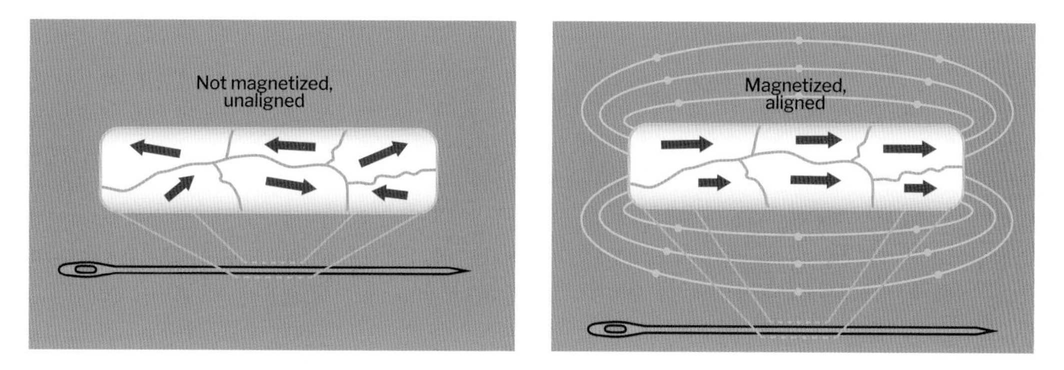

Domains within a needle align when it is magnetized.

The Northern lights are a result of Earth's magnetic field.

of the planet, but near the poles the field is weaker. Here, some particles of solar wind break through. Collisions between solar particles and the molecules of gas in the atmosphere produce flashes of light which we see as the northern or southern lights.

MAGNETISM FOR MOVEMENT

Particles with an electric charge, whether positive (protons) or negative (electrons) create a magnetic field as they move. As electrons flow through a wire, a magnetic field is set up around the wire. The direction of the current determines the direction of the magnetic field, and the effect is strongest closest to the wire. The strength of the magnetic field increases with the current. This allows the creation of electromagnets. A simple electromagnet can be made by winding wire around an iron nail and passing an electric current through the wire. Scaling this up produces hugely powerful electromagnets of the type used to pick up cars in wrecking yards. These are not permanent magnets: when the current is turned off, the magnetic field disappears.

The interaction of an electric current and magnetism can be exploited to make an electric motor. If a loop of wire has a current passed through it, a magnetic field is set up around the wire, all along

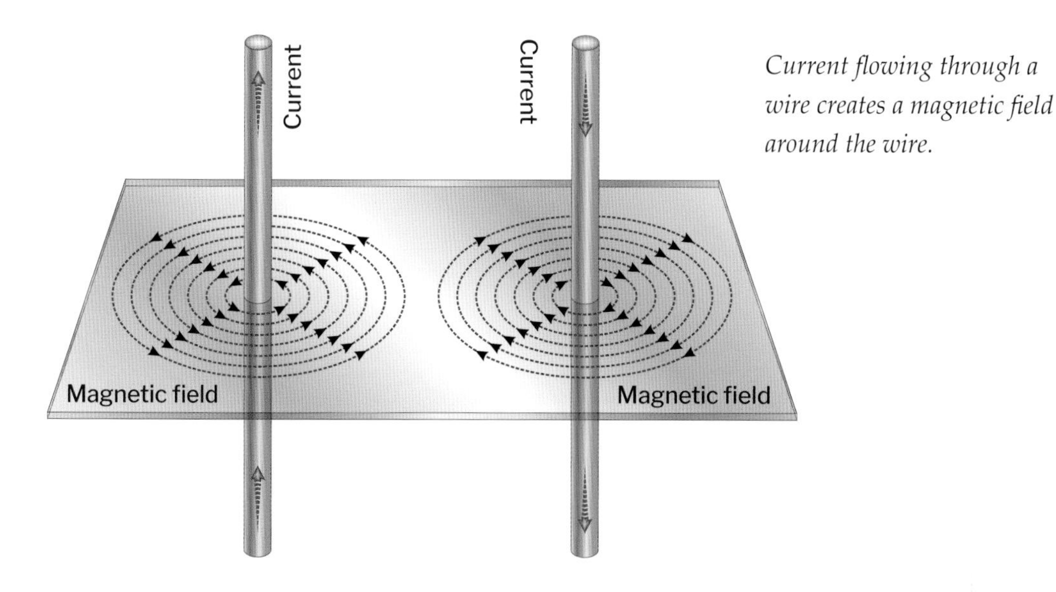

Current flowing through a wire creates a magnetic field around the wire.

Floating with magnetism

Maglev trains are simultaneously pushed and pulled along by magnetic fields which also keep them raised above a track. Powerful magnets line the underside of the train and metal coils line the guideway. An electric current passed through the coils creates a magnetic field. Switching the direction of the current, magnets in front of the train pull it and those behind push it forwards. The train is kept floating above the guideway by magnetic repulsion. As there are no wheels rolling over tracks, there is very little friction to slow the train down, and none of the noise and vibration associated with regular trains.

its length. If the loop is then placed in another magnetic field, at right angles to the direction of the current, the result is to turn the loop as the external magnetic field interacts with that created by the current in the wire. By reversing the current at just the right interval, the loop is kept turning in the same direction – and that provides the movement for the motor.

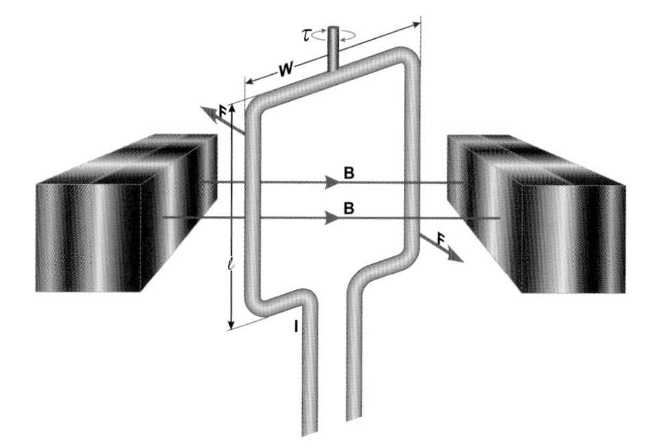

A motor is produced by passing an electric current through a wire loop perpendicular to a magnetic field.

29

DNA

The book of life

The recognition that chromosomes act like a recipe or series of instructions for making an organism raised the prospect of reading those instructions directly and perhaps even altering them to change an organism in carefully targeted ways. The first step was to discover the structure of DNA, the molecule that makes up chromosomes and so holds the instructions.

UNRAVELLING THE THREAD

DNA (deoxyribonucleic acid) was first discovered in 1869 by Friedrich Miescher. He found it in the nuclei of white blood cells, so named it 'nuclein'. Although he felt it was important, it took more than 50 years for other scientists to realize just how important it is and work out its function. In 1944, Canadian-American physician Oswald Avery found that it's the 'transforming principle' that enabled the bacteria he was studying to move characteristics between strains. Six years later, in 1950, Erwin Chargaff published his findings: that DNA differs between species, and a DNA molecule always contains equal numbers of the bases (see box opposite) guanine and cytosine, and equal numbers of adenine and thymine.

A race ensued to work out the structure of the DNA molecule, and how it could convey so much information. It was won by Francis Crick and James Watson in 1953, working from X-ray photos taken by Rosalind Franklin the previous year. It emerged that the particular sequence of the bases is the way the instructions to make an organism are 'coded' into DNA.

The longest human chromosome, chromosome 1, contains more than 240 million base pairs. These are divided into areas providing 3,000 genes and much additional material between them. Each gene holds the code for constructing one of the proteins essential to building and running living bodies. Additional material contains information such as which genes are active and which not.

MAPPING THE GENOME

The first step in working with DNA was to 'sequence' the genomes of organisms. This means listing the base pairs on each chromosome. Between 1990 and 2003, the Human Genome Project mapped

The double helix

DNA is a very long molecule that looks rather like a twisted ladder. It's a double helix, comprising two strands of a sugar (deoxyribose) and phosphate molecules, linked by pairs of bases that make the 'rungs'. The bases are molecules made of carbon, nitrogen, hydrogen and oxygen, differently arranged to make four different chemicals: adenine, cytosine, guanine and thymine. On one side, each base is chemically bonded to the sugar-phosphate backbone. On the other side a strong hydrogen bond links it to its corresponding base: adenine with thymine and cytosine with guanine.

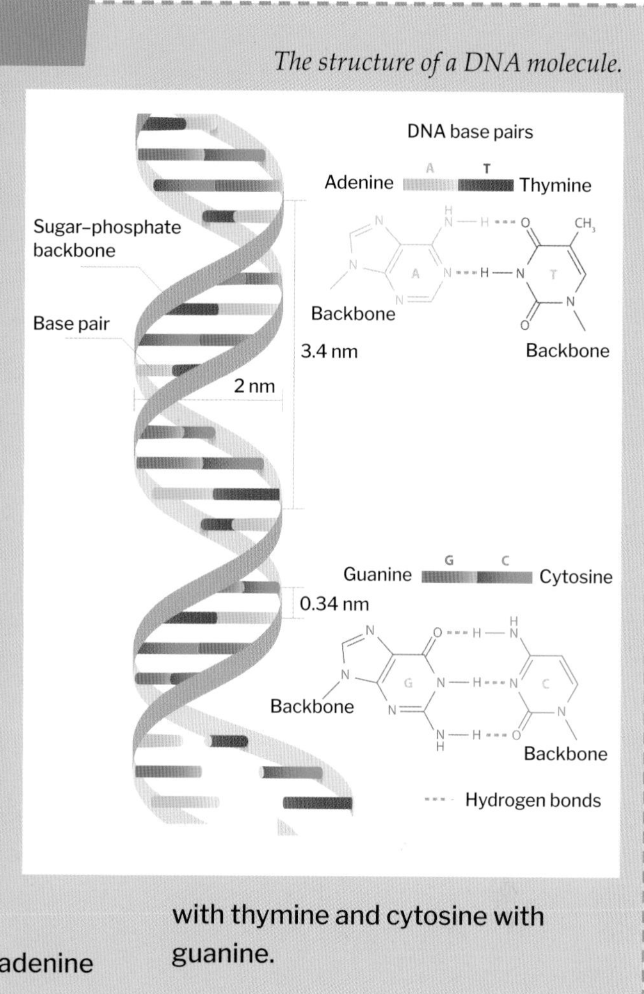

The structure of a DNA molecule.

the human genome, and the genomes of several other organisms along the way.

Although difficult, sequencing is a straightforward chemical task – we can look at the molecule and see where the base pairs occur. Finding the functions of genes is more complicated, but it's necessary if we are to do anything useful. Genes aren't labelled with their functions. These can only be identified by laboriously comparing the chromosomes of organisms with different features and finding where the genetic differences are. Research has focused on identifying the human genes implicated in disease and disorder, enabling genetic medicine to tackle these conditions.

MAKING CHANGES

Understanding genomes enables us to make changes to crops and farm animals, identify people who might be susceptible to genetic disease, and help people have healthy babies. Some of these activities

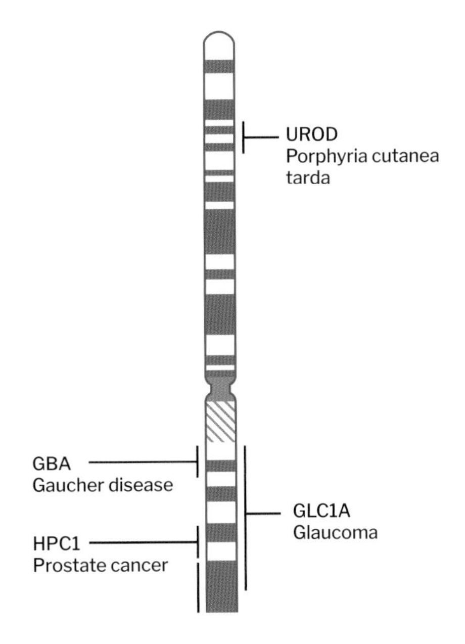

Part of chromosome 1, showing areas responsible for susceptibility to certain diseases.

fall under the broad label of genetic engineering. Of course, humans had been changing the genetic makeup of plants and animals for millennia by selective breeding, but genetic engineering meant we could do it much more quickly and precisely.

Some of the uses of genetic engineering have been rather trivial – such as making day-glo pet fish. Many more have been life-altering or have large-scale social or economic implications or impact. Insulin, needed by people with diabetes, used to be taken from animals but is now manufactured in giant vats by bacteria that have been genetically modified to produce it. Genetically modified organisms (or GMOs), include plants that are drought-, frost-, or pest-resistant, and crops that have enhanced nutritional value (such as extra vitamins).

There has been considerable public concern and outcry about GMOs and whether their safety can be guaranteed either in the food chain or if they spread freely in the wild. There is controversy, too, over companies that develop crops that are resistant to weed-killers so that the organization can sell not only seed but the targeted weed-killer that goes with it, tying farmers into further financial commitments. Different parts of the world have taken different stances on these issues. More ambitious programmes have made pigs with human-compatible organs for transplant, or attempted to make a cow that will produce milk more like human breast milk.

HUMAN GENES

Although some people have talked about using genetic engineering to eliminate genetic diseases or to produce 'designer babies' (children with desired characteristics, such as a particular eye colour or abilities), this is not currently legal. Genetic screening, though, gives couples with a family history of genetic disease a chance to have a healthy baby. They can opt to use IVF (in-vitro fertilization), having their fertilized eggs screened for the genes that will cause damage and only implanting healthy embryos. This doesn't involve making any changes to chromosomes or genes, and it doesn't prevent a child having some other genetic disorder that hasn't been screened for.

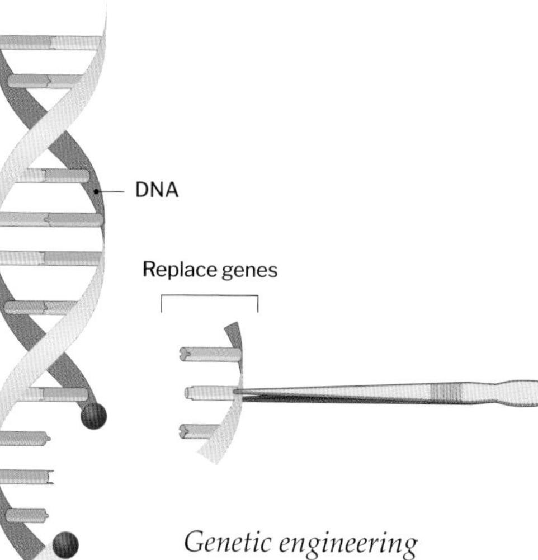

Genetic engineering techniques such as CRISPR replace a small portion of DNA to manipulate features of an organism.

MOVING ON – CRISPR

Gene editing has been restricted to changing crops, bacteria and farm animals, and was expensive until the development in 2012 of a new technique called CRISPR-Cas9. CRISPR is quick and cheap, making revolutionary change possible. It exploits a natural process: bacteria protect themselves against disease by copying and storing bits of viral RNA. When they encounter the virus again, the bacteria can recognize it and snip it into pieces. The CRISPR technique uses the same mechanism, sometimes described as genetic 'scissors', to cut out part of a chromosome and patch a healthy gene in its place.

Before CRISPR, genetic modification acted at the level of eggs, sperm and fertilized eggs. It changed the genes of a new organism before it developed. With CRISPR, it's possible to change cells in a fully-grown body. The first disease to be treated with a CRISPR-based treatment was sickle-cell disease, which causes red blood cells to be distorted into a shape inefficient at carrying oxygen. So far, it remains illegal to alter human fertilized eggs intended for pregnancy, but one doctor in China claims to have edited eggs to give twin babies protection against HIV in 2018.

Germ-line and somatic cells

Somatic cells are body cells. They are all the cells in your body that are already differentiated as muscle, bone, blood and so on. They can divide to produce more cells of the same type, and some (such as stem cells in the bone marrow) can produce a range of different cells. But these cells remain in your body, and any changes to them are not passed on to any future children you might have.

Germ-line cells, on the other hand, are the egg and sperm cells involved in reproduction. These cells provide the DNA for the next generation. Any changes to the DNA of these cells will be passed on, and will be incorporated into the eggs or sperm that child produces.

30

The matter of matter

The standard model

Modern physics is based on the so-called 'standard model' of matter developed in the 1970s. It identifies a limited number of fundamental particles that are the building blocks of all matter and mediate forces acting on matter.

MAKING MATTER

So far, we've seen how atoms are made of protons and neutrons forming the nucleus and electrons orbiting it. It turns out even the protons and neutrons can be further divided.

All normal matter is made of two types of basic building blocks: quarks (pronounced 'quorks' to rhyme with 'storks') and leptons. There are six types of each. The quarks are grouped in pairs or 'generations', named up, down (first generation), top, bottom (second generation), and strangeness and charm (third generation). Protons and neutrons are each made up of three quarks. An electron is one of the six types of leptons. The leptons are also paired: the electron and electron neutrino, the muon and muon neutrino, and the tau and tau neutrino.

QUARKS

Quarks were proposed independently by Murray Gell-Mann and George Zweig in 1964. At the time, physicists were using accelerators to smash particles together at very high speed, then looking at the results. Gell-Mann was trying to organize and explain the proliferation of different particles that were turning up from collisions. He realized that the behaviour of many of them could be explained if they were made up of even smaller particles. He called these hypothetical particles quarks.

Protons and neutrons are of equal mass, yet protons have a positive charge equal to the negative charge on an electron, and neutrons have no charge. Gell-Mann explained this by dividing protons and neutrons into three quarks each, and giving the quarks positive and negative charges. But unlike any other known particle, quarks have a fraction of the charge on an electron, the 'elementary charge', e. This suggestion was considered ridiculous at the time, but it turned out to be correct, with evidence for the existence of quarks emerging in

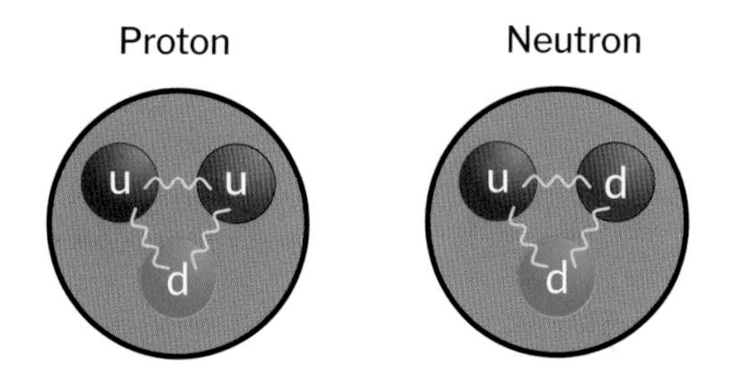

Quarks combining in protons and neutrons; the wiggly lines are gluons.

the early 1970s. The charge on a quark is either –1/3e or +2/3 e. Up quarks have a charge of +2/3 e and down quarks have a charge of –1/3 e. By combining an up and two downs, a neutron has no overall charge. By combining two ups and a down, a proton has a charge of +1 e.

CHANGING QUARKS

Most quarks are the stable, lower energy (first generation), 'up' and 'down' quarks. The other four types are high-energy quarks that quickly decay into up and down quarks. Charm and top are both up-type quarks that decay into up quarks. Strangeness and bottom decay into down quarks. In addition, a down-type quark can decay into its corresponding up-quark by emitting a W^- boson (and vice versa – an up-type quark can interact with a W boson to become the corresponding down-type quark). In this way, W bosons mediate radioactive decay. If a down quark in a neutron emits a W^- boson, it becomes a proton: the change from down quark to up quark comes with a change of charge from –2/3 e to +1/3 e. The charge is lost in the form of an electron (-1 e). This emerges when the W boson decays into an electron and an electron neutrino.

LEPTONS

Leptons are more straightforward as they exist on their own rather than combining to make other particles. (Quarks are never found alone.) The electron, muon and tau all have an electrical charge and significant mass, while their corresponding neutrinos have very little mass and no charge.

Neutrinos are produced whenever atomic nuclei come together or break apart. They're elusive particles that can wink in and out of existence, flip between types, or even combine types and masses (an example of quantum superposition, see page 195). A neutrino has less than a millionth the mass of an electron and can pass straight through most matter, including Earth, which makes them very difficult to detect. Predicted in 1930, they weren't found until 1956. Even so, they are the second most numerous particles in the universe after photons.

FEEL THE FORCES

The particles that make up matter, quarks and leptons, are acted upon by forces, and these are mediated by another group of particles, called bosons. Photons mediate the electromagnetic force which holds atoms and molecules together. The weak

This type of diagram, called a Feynman diagram, is a kind of narrative that shows the behaviour of subatomic particles. Passing time is vertical; the bottom of the diagram is the earliest stage. Here, we start with a neutron, with up, down, down quarks (udd). At the angle in the diagram, the neutron interacts with a W⁻ boson and one of the down quarks loses an electron and an electron neutrino, becoming an up quark (udu). This changes the neutron into a proton. It alters the atomic number of the atom, so changing it to a different element – the hallmark of radioactive decay.

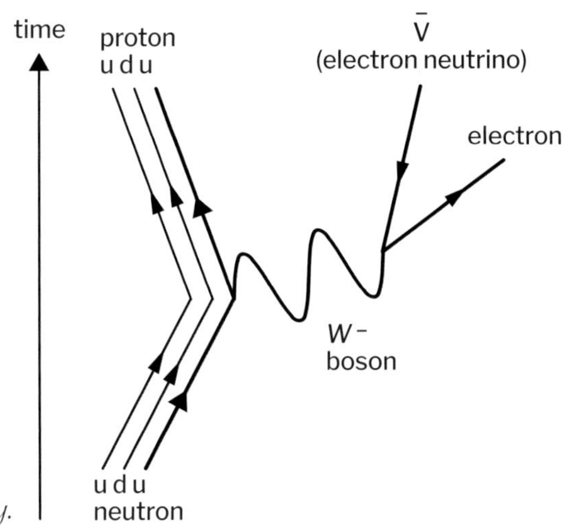

nuclear force, essential to radioactive beta decay, is mediated by W and Z bosons. Gluons produce the strong nuclear force, holding quarks together in the protons and neutrons of atoms. Gravity is the one that hasn't been fully explained. It might be mediated by gravitons, but they haven't been found yet.

AND AN EXTRA...

In addition, there's an extra boson, called the Higgs boson. First proposed in 1967 and detected in 2012, it's thought to give

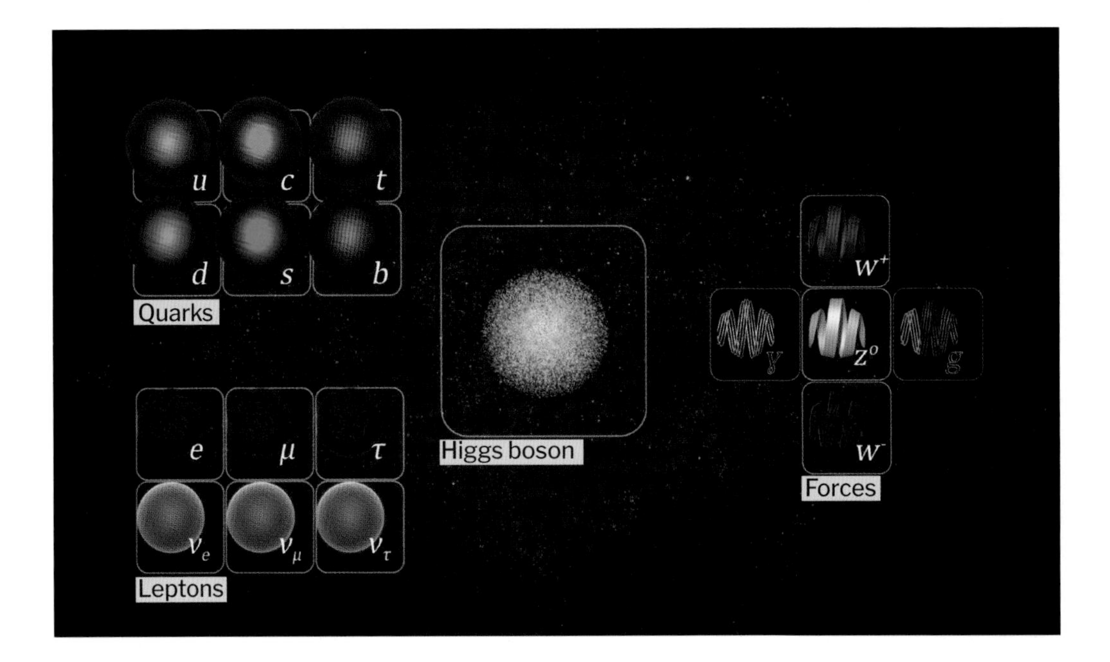

The Standard Model – leptons, quarks and bosons.

mass to particles. In the 1970s, scientists realized they could combine the weak nuclear force and the electromagnetic force in a single 'electroweak' force. On the whole, the mathematics worked, but equations came out with the bosons involved having zero mass. While this is true for photons, the W and Z bosons do have mass. A new suggestion answered that problem, but needed a new particle. Here it is: as the universe cooled after the Big Bang, a new field, called the Higgs field, spontaneously appeared, everywhere. Interaction with this field gives particles mass – and the field is mediated by the Higgs boson. The more a particle interacts with the field, the more mass it has. Photons don't interact with it at all, and so have no mass. Other particles interact to differing degrees, and so have different masses.

Simulation of a particle collision creating a Higgs boson.

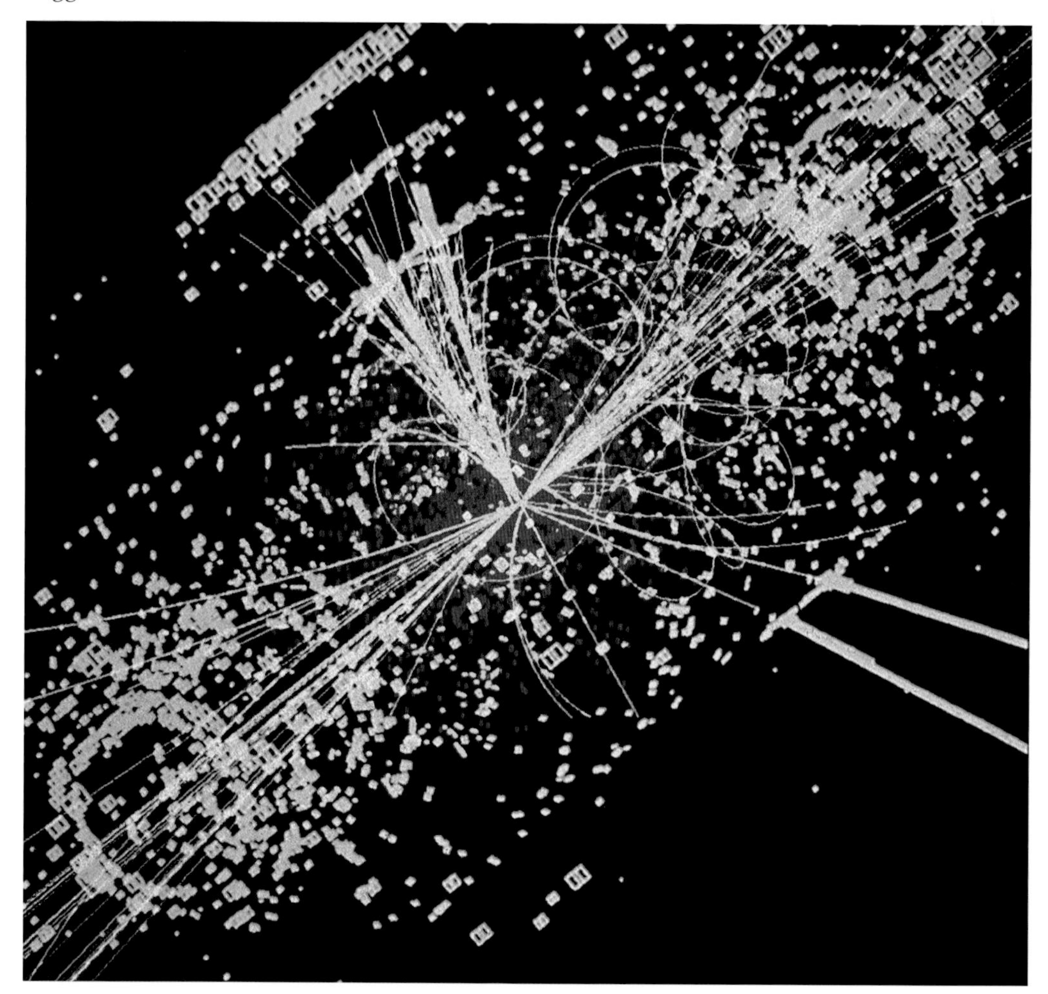

31

Keep it clean

Antisepsis

If you have an operation or an injury today, you'd be unlikely to die from an infected wound, but in the past it was a common outcome. Before people understood that germs cause disease and infection (see page 75), they had no reliable means of preventing infection. Since the development of antiseptics in the 1860s, surgery and accidental injuries have become much less dangerous.

A FIRST LINE OF DEFENCE

There are millions of microbes all around us in the air, water, soil and on the surface of our bodies. When the skin is broken, microbes can easily get inside the body and some of these are harmful. Harmful microbes are commonly called germs. In the warm, wet environment inside the body, they reproduce very rapidly and can destroy tissues or cause illness. Although our bodies have defences against invasive microbes, these can be overwhelmed, especially if someone is weakened by illness or trauma. An infected wound becomes septic, producing pus, becoming inflamed and painful, and leading to the death of tissue around it. The infection can spread, causing a whole limb to become gangrenous, and can lead to death.

Once an infection has taken hold, antibiotics can now be used to treat it (see page 136), but a good way to prevent infection setting in is to use antiseptics to kill germs as they arrive.

USE WITHOUT UNDERSTANDING

People discovered long ago that some substances seem to protect against infection. In Ancient Greece, the physician Hippocrates (460–377 BC) recommended using wine and vinegar on wounds. These would help to destroy germs. In Ancient Persia (now Iran), drinking water was kept in copper pots, which have antiseptic properties, and in both Peru and South East Asia, balsam (a resin from plants) was used to protect wounds. It was used in Europe from the Middle Ages until the 19th century.

The great French military surgeon Ambroise Paré discovered on a battlefield in 1536 that the treatment of terrible injuries with oil of roses, egg yolk and

Ambroise Paré.

turpentine had better outcomes than the conventional and brutal practice of cauterizing with hot iron or boiling oil. His patients were lucky that he had run out of boiling oil and tried something else. Not only were they less likely to die of shock, their wounds were less likely to become fatally infected as the mixture provided an environment hostile to germs.

In the 18th century, Sir John Pringle (1707–82) studied a series of substances with antiseptic properties. Mercury chloride and then iodine were soon added to the list of useful applications, but none of these was good enough to protect much more than a minor injury. The death rate following amputations ran at nearly 50 per cent, as there was simply no way of preventing serious infection.

NEXT TO GODLINESS?

The first rule for preventing infection today is cleanliness. This is now so obvious that it is horrifying to realize that only 200 years ago doctors would go from one patient to the next without

Antiseptic or disinfectant?

Antiseptics are used on and near living tissue to prevent infection. Disinfectants are used on non-living surfaces. They can be harsher as they aren't used on living bodies. Bleach is an effective disinfectant because it kills many types of cells – and that includes human body cells.

washing their hands or cleaning their instruments. Although some physicians in the distant past were well aware that hygiene seemed to improve outcomes for patients, it was clearly forgotten somewhere along the way. Surgeons of the 19th century wore their everyday clothes, perhaps covered with an apron or gown that accumulated blood, pus and gore during the day as they worked. Their hands and instruments would carry germs from one patient to another. They did not clean them, even if they had just come from dissecting an infected corpse. It's hardly surprising patients were dying.

In 19th-century Vienna, women who went into hospital to have their babies stood a good chance of dying – nearly a 20 per cent chance on some wards. The cause was puerperal fever, a bacterial infection of the uterus caused by streptococcus. It ripped through maternity wards leaving a swathe of dead new mothers. Hungarian physician Ignaz Semmelweis, greatly troubled by the figures, began to wash his hands with soap and chlorine and to ask his students and colleagues to do the same in 1847. The idea came to him after a colleague cut his finger during an autopsy and died from an infection that looked similar to puerperal fever. Just washing their hands cut mortality from 18 per cent to around 1 per cent. Semmelweis was not the first to think of this, though he was the most famous. Oliver Wendell Holmes in America and Alexander Gordon in Scotland had said much the same but didn't have a hospital to run a trial. Even so, the advice had little impact as no one understood how it worked and many scoffed at it. It didn't fit the contemporary model of infection as being caused by an imbalance of 'humours' or 'bad air'.

SOLUTION FROM A SMELL

The breakthrough came in 1865 when the English surgeon Joseph Lister read Pasteur's work on germ theory. Suspecting that microbes from the air were causing infection in wounds, he looked for a way of killing them or preventing them multiplying. At the time, carbolic acid (phenol) was used to treat sewage to prevent it smelling. It did this by killing the microbes that caused the rotting sewage to smell; Lister tried using carbolic acid as an antiseptic. His first patient, in 1867, was a seven-year-old boy who had a compound fracture after being hit by a horse-drawn cart. At the time, compound fractures – those in which

Joseph Lister.

the broken bone cuts through the skin – were almost always fatal. He sprayed the injury and surrounding area with carbolic acid when operating on the leg and laid bandages soaked in carbolic acid over the boy's wound. The boy escaped infection and survived. Lister went on to perform many operations using a spray of carbolic acid, then covering wounds with antiseptic-drenched dressings. This way, he killed any bacteria present and prevented more colonizing the wound.

A CLEAN CUT

The development of anaesthetic had actually made deaths from surgery worse rather than better. With immobilized patients, surgeons could do more complicated operations and could open up the main part of the body, which had previously been impossible. But by doing this, they allowed bacteria in. Only with the development of antisepsis did lifesaving abdominal operations become truly possible.

32

Kill the germs

Antibiotics

Antisepsis is a method of preventing bacteria and other microbes starting an infection. What if the infection is already there? Then you need antibiotics – as long as the infection is bacterial.

A NARROW BRUSH

Antisepsis is a broad-brush technique. It works by creating a hostile environment that destroys the cells of bacteria and other biological agents of infection. But antibiotics are more subtle and more precise. They work only against bacteria, and won't resolve an infection caused by viruses, fungi, parasites, or anything else. There are many types of antibiotics and they each work against different groups of bacteria. Some are 'broad-spectrum' antibiotics that kill a wide range of bacteria, including those that are helpful to us (such as gut bacteria involved in digestion). Others only work against a narrow range or a few specific bacteria. Specific antibiotics are prescribed for each particular bacterial infection.

A MISTAKE WITH MOULD

The first antibiotic identified was penicillin. It was discovered in 1928 by Alexander Fleming. Fleming had been researching the bacterium *Staphylococcus*, which causes boils and sore throats, and before going on holiday he piled up his petri dishes with cultures of the bacteria without cleaning them. When he returned, he found clear circles on one of his plates where the bacteria had been killed. He investigated and found that a mould had secreted a juice of some kind. He and his assistants set about trying to isolate the ingredient that killed the bacteria. By the end of his research, Fleming had found that the mould *Penicillium notatum* produces a substance that is quite unstable and difficult to work with, but that kills some types of bacteria. It was not yet a medicine, and looked as though it might be useful in microbiology labs to separate different kinds of bacteria.

Fast-forward ten years. On the brink of World War II, Howard Florey and Ernst Chain began work on producing and extracting the juice that *Penicillium* makes. They processed 500 litres (110 gallons) a week of liquid, growing the fungus in every receptacle available, including bathtubs and sinks. Their

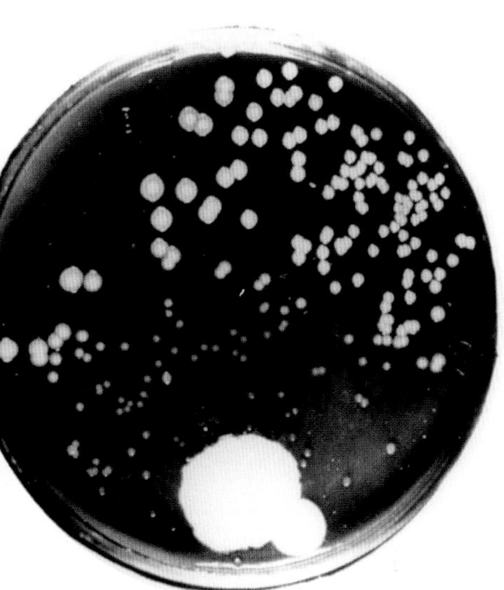

Left: Alexander Fleming.
Above: Fleming grew his bacteria and antibacteria on plates like this.

work bore fruit: in 1940, they showed that the extract could protect mice against infection with *Staphylococcus*. The following year, they tried penicillin on the first human patient, a police officer who had scratched his face while pruning roses and developed an infection. Injected with penicillin, his symptoms cleared up – but the supply of penicillin ran out and a few days later he died. Florey and Chain had the result they needed, and penicillin began to be used successfully on other patients.

MORE AND MORE

Since the discovery of penicillin, more antibiotics have been found and developed as medications. Many are made by fungi, often found in soil, but they can be found in other environments. Some come from algae, lichens, plants and even other bacteria. In nature, antibacterial chemicals are produced by living organisms to protect themselves against attack by bacteria. The task is to find, identify and reproduce these chemicals.

A GOLDEN ERA

Antibiotics revolutionized medicine. Until the 1940s, there was no way of combatting infections caused by bacteria invading wounds. Accidental injuries and surgical wounds were both vulnerable to infections that could quickly become deadly. Other infections caused by bacteria, from tuberculosis to urinary tract infections, are also treated with antibiotics. Before antibiotics, there were no treatments for these conditions,

Mass production of penicillin during World War II.

main cause of death in developed countries changed from infection to non-infectious diseases such as heart disease and cancer. People no longer died from a tooth abscess, a bladder infection, or a scratch from a rose bush. From the 1950s to the 1970s, new groups of antibacterials were discovered regularly, and medicine could combat almost any infection.

or rheumatic fever, bacterial pneumonia and a host of other illnesses. Life expectancy rose in countries that used antibiotics. In the USA, the proportion of the population who were elderly rose from 4 per cent to 13 per cent. The

TURNING BACK THE CLOCK

But the golden years are over. Antibiotics were so successful in combatting infection that they were used indiscriminately and unnecessarily. They were given to people with colds – a viral infection against

How antibiotics work

Antibiotics use different strategies to attack bacteria. They can kill the micro-organisms directly, or prevent them growing or reproducing. Some stop bacteria making the proteins they need. Some stop them making DNA or break up their DNA. Others disrupt the cell wall or membrane so that the bacterium can't control what goes in and out of the cell, or disrupt the chemical reactions that take place in the cell. They don't harm cells in the person or animal that takes the antibiotic because their cells are very different from bacterial cells.

which they are useless. They were given in huge quantities to farm animals to make them grow more quickly and to prevent them getting infections. Bacteria reproduce quickly and that means they can evolve quickly. Each reproduction gives them the chance to mutate, to gain new characteristics, including resistance to the antibiotics used against them. Over recent decades, more and more antibiotic medicines have become ineffective as the bacteria they once killed have upped their game: they are now resistant, and the drugs don't work. The first bacterium to become resistant was a *Staphylococcus aurea* – the very bacterium Fleming worked on in the 1920s. MRSA (multi-resistant *Staphylococcus aurea*) is now resistant not just to penicillin but to almost anything we can throw at it. People are, once again, dying in hospitals of infections that could be treated in the 20th century.

GROWING RESISTANCE

Bacteria are simple, single-celled organisms. Some can pass genetic material between species. This means that if one type of bacterium is resistant to an antibiotic, it might pass that resistance on to other bacteria that were previously killed by the antibiotic. Antibiotic-resistant bacteria will survive where their weaker colleagues are killed. They then become the dominant strain: it's evolution in action.

Bacteria have several mechanisms to resist antibiotics.

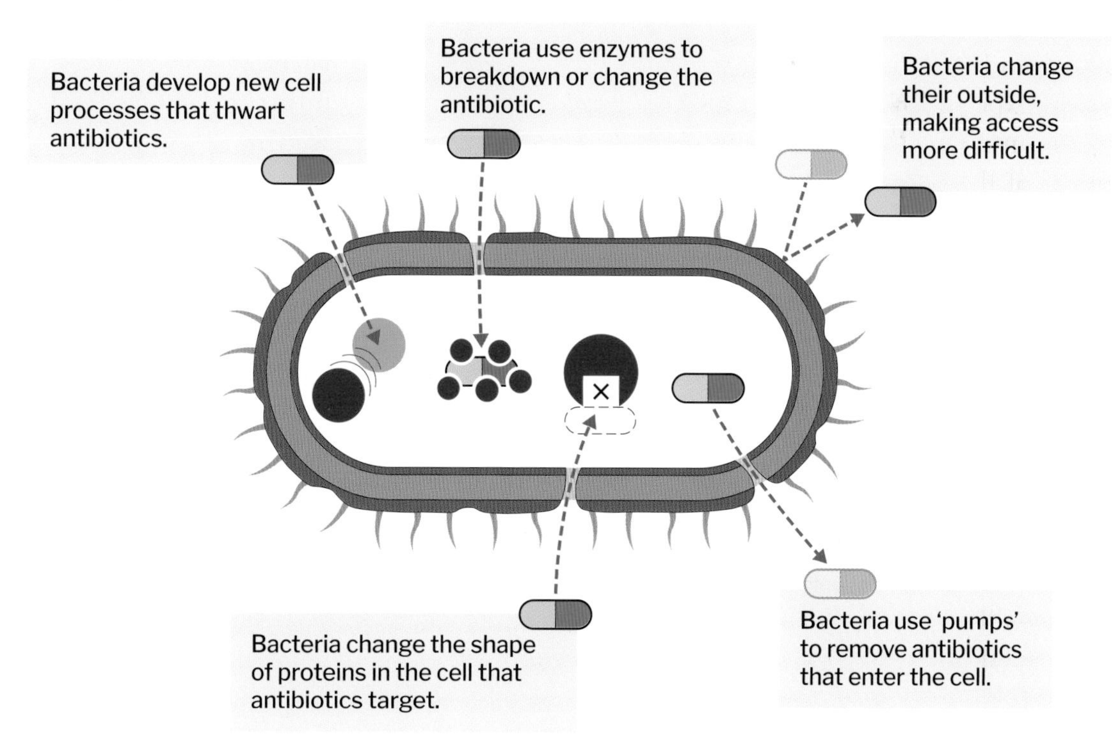

33

You can't lose

The conservation of matter

The atoms of oxygen you breathe in today were present when Earth formed 4.5 billion years ago. They will still be here in another two billion years' time. And that's true of all the atoms that have been built into your body, and that you take in as food.

PHYSICAL CHANGES

Matter can't be created or destroyed, but it can and does change its form, changing physically and chemically. A physical change does not make any change to the individual atoms and molecules, and is reversible.

Water is an example of a substance that frequently undergoes physical change. Water can exist in three forms on Earth: as a solid (ice), as liquid water and as a gas (water vapour). It changes between these forms when it is heated or cooled. The water molecules are the same in all cases, but they are differently arranged and move differently in the three states. In ice, they are held in a crystal lattice and can only vibrate. In water they can move around, so water flows. In a gas, they can move very freely and are much further apart. One gram of water vapour takes up far more space (has a greater volume) than a gram of water or ice because the water molecules are more widely separated.

No molecules are added, removed or changed when water makes the physical change between states. A quantity of water that is frozen into ice has the same mass before and after freezing and will have the same mass if melted or vaporized.

CHEMICAL CHANGES

Chemical change does affect the individual atoms and molecules. Chemical change alters the starting substances completely. Examples are burning oil or rusting iron. The substances produced by the change have different properties from the substances at the start of the process.

The whole of life and most non-living processes take place through chemical reactions. These take apart the molecules of the reactants (the starting chemicals) and make new bonds between atoms to make the products of the reaction. When you eat food, it is chemically unpicked in your gut during digestion and your

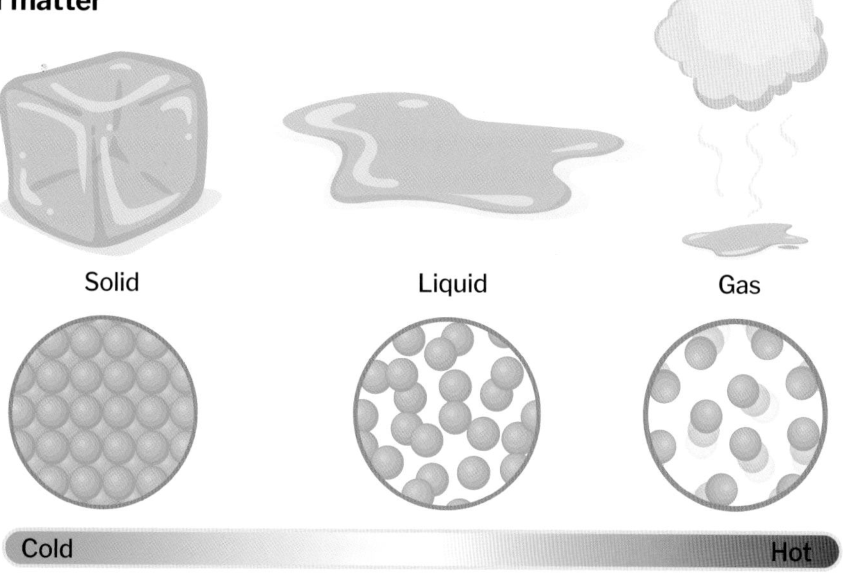

Different states of water.

body makes use of the chemicals it needs, building them into new molecules, and expels the rest. The mass of food you eat is the same as the mass of chemicals stored in or used by your body plus the mass of those excreted. Similarly, when the surface of a rock dissolves in acid rain, it doesn't just disappear. Particles of rock and dissolved chemicals are washed away, and eventually arrive in a river or the sea. Some chemicals might escape into the atmosphere as gas. The mass of rock that has been removed is the same as the total mass of tiny particles of sediment, of dissolved minerals, and of released gas that result from the reaction.

Chemical reactions are represented by equations which show how many atoms or molecules of each reactant are needed and how many atoms or molecules of each product are made in the reaction. The equation must balance, because no matter can be created or destroyed. This is the reaction for photosynthesis, the process by which plants make glucose ($C_6H_{12}O_6$) and oxygen (O_2) from carbon dioxide (CO_2) and water (H_2O):

$$6CO_2 + 6H_2O \rightarrow C_6H_{12}O_6 + 6O_2$$

On each side of the equation, there are six carbon atoms, 12 hydrogen atoms (first present as six molecules that each have two hydrogen atoms) and 18 oxygen atoms (12 in the carbon dioxide molecules and six in the water molecules). This can be scaled up to any quantity: the equation shows the ratios you would need in terms of the number of atoms.

STARTING OUT WITH MATTER

The first atoms to form in the universe were hydrogen and then helium. The other

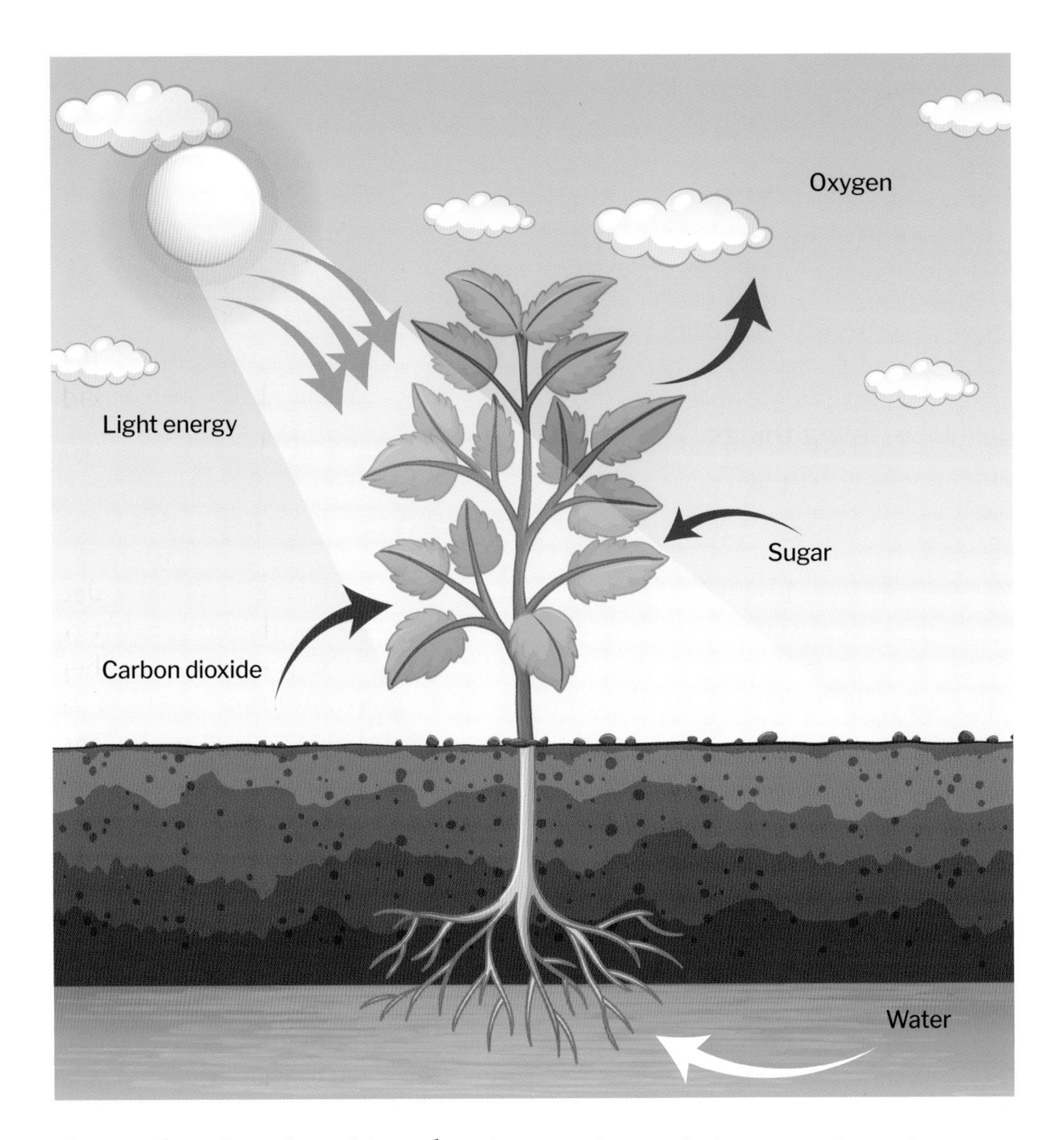

By photosynthesis, a green plant makes glucose from water and carbon dioxide, using the energy of sunlight.

elements have been forged from these in the hearts of stars and in supernovae – the massive explosions in which some large stars end their lives. Those atoms have been recycled for billions of years, forming compounds, and then being freed from those compounds to make other compounds again and again. Our planet, being only 4.5 billion years old, uses atoms that were previously in some other solar system, possibly in another, long dead, planet. And after Earth is no more, they will go on to survive elsewhere in the universe.

34

Everything in order

Classifying organisms

People have a strong urge to put things into groups, to make distinctions based on similarities and differences. The very existence of words like 'plant' and 'animal' shows we distinguish between broad groups of living things. There is no generic plant or animal. We can only illustrate these terms by giving an example.

An interest in formalizing such distinctions started early, and led eventually to the two-part biological naming system we now use for all living organisms.

SEEING DIFFERENCE

The Greek philosopher Aristotle (384–322 BC) made the first known attempt at classifying living things. He started by dividing plants from animals, and then grouped animals based on differences between them. (His work on plants has been lost.) He divided animals first into those with red blood and those without, which coincides with our distinction between vertebrates (those with a backbone) and invertebrates (those without, as vertebrates have red blood). Next, he divided each of those groups into five.

Soon after, Aristotle's pupil Theophrastus (371–287 BC) classified plants as trees, shrubs, herbaceous perennials and herbs. He subdivided them, describing their methods of reproduction, their sizes, where they grew, and their practical uses as foods, herbs and so on. Aristotle and Theophrastus based their classifications on features that could be observed rather than innate properties; a plant was a tree if it was large and woody, a shrub if it was shorter.

A LONG RUN

These early classifications were used for around 2,000 years. Then, in the 16th and 17th centuries, Europeans began travelling further afield and collecting biological specimens. They used new methods to classify the novel organisms they found. But with several systems being used, the result was confusion rather than clarity. The same plant might be given different names in different systems.

The Swedish botanist Carl Linnaeus overhauled and rationalized the mix of

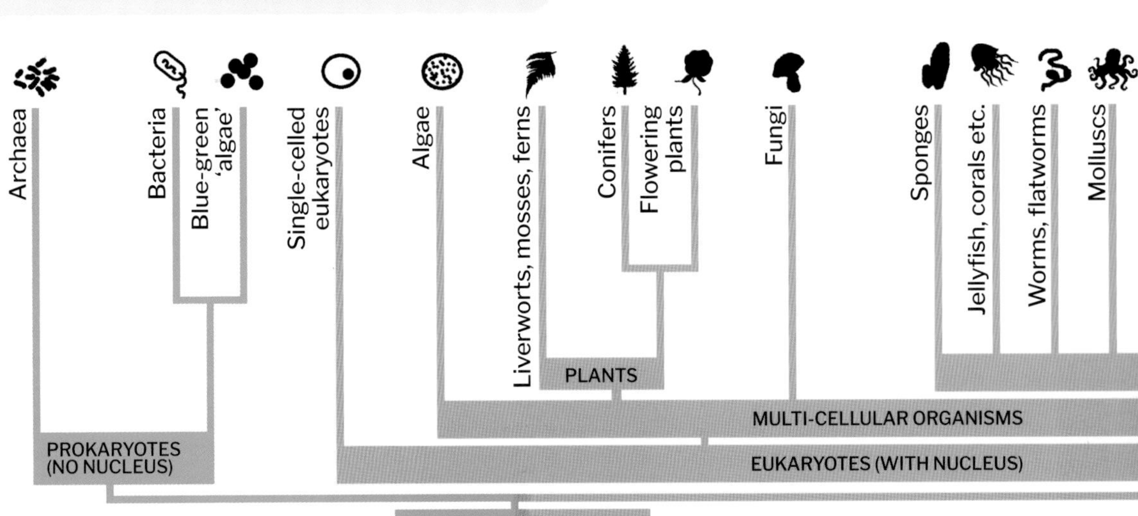

systems, publishing his own classification in 1735. He divided all living organisms between two 'kingdoms', plants and animals, and had a third kingdom, minerals, for non-living things. Then he split each kingdom into five further levels: class, order, family, genus and species. Similar but different examples within a species could also be divided by variety. Each organism was given a two-part name in Latin, showing first its genus and then its species. Linnaeus had six classes of animal (mammal, bird, amphibian, fish, insect, worm) and 24 classes of plant, the first 13 based entirely on how many stamens the flowers had. Number of stamens is a clearly visible difference between flowers, but is no longer considered to be as important as Linnaeus made it. His final class of

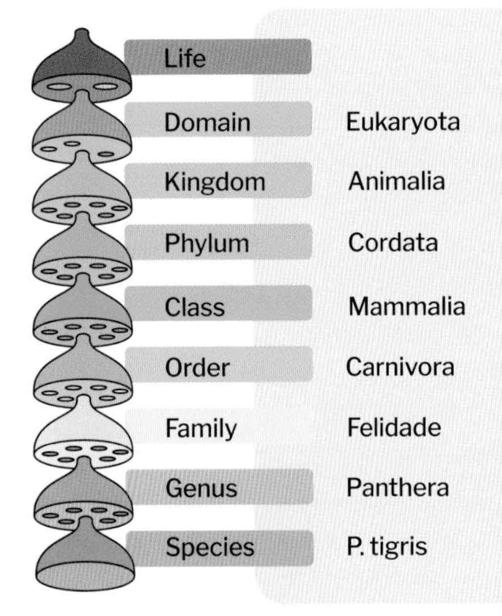

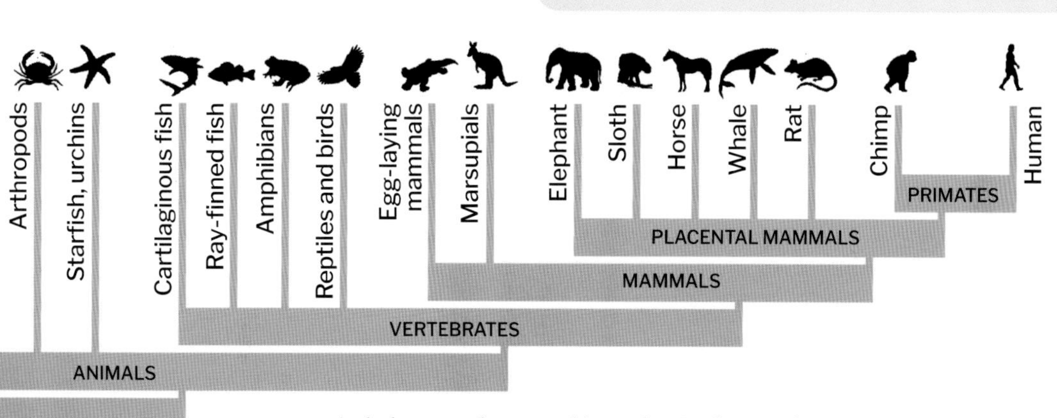

A cladogram shows no hierarchy. It shows where organisms diverged genetically from their predecessors.

plants included all ferns, algae and fungi. (Fungi are not now considered plants.)

With some modification, Linnaeus's system is still used for plants and animals (but not minerals). New organisms can easily be added, sometimes as extra species in a genus and sometimes needing an entirely new genus.

Linnaeus tried to base his classification on intrinsic features of the plants and animals he categorized. Even this, though, relied on people perceiving some features as more important than others. It would be a poor classification system that grouped animals by colour, for instance. That would put different species of bee as brown animals and yellow animals, when they are clearly closer to one another than to, say, brown bears and yellow canaries. Linnaeus and those following him tried to identify meaningful similarities to help them group animals and plants. By this means, a bat clearly shares more features with other mammals than with birds and so is classed as a mammal even though it flies.

UPDATED

The Linnaean system is still used but has been considerably updated over nearly three centuries. A new level above kingdom was added in the late 20th century dividing living things into archaea, bacteria and eukaryotes. Archaea and bacteria are both simple life forms with a single prokaryotic cell (see page 48), while all organisms with more sophisticated eukaryotic cells, ranging from single-celled algae to animals and plants with billions of cells, are eukaryotes.

Modern cladistics tries to root classification in evolutionary relationships between organisms, finding common ancestors that link them. These relationships can be tested with DNA analysis, but Linnaeus had no access to any method of objectively measuring the relatedness of organisms. Cladistics has led to further changes in classification, with some biologists now rejecting the kingdoms that eukaryotes are divided into because they are not monophyletic – all the members of the kingdom don't come from a single common ancestral group.

35

Cells within cells

Endosymbiosis

At some point in the last 4.5 billion years, the first single-celled organisms evolved. These were prokaryotes – organisms with a single, simple cell. We shouldn't be too dismissive of them on account of their simplicity, though. They were the only living things for at least a billion years.

Without them, we wouldn't be here – and not only because all today's organisms evolved from them, but because they have left their traces in nearly every cell of our bodies.

FROM PROKARYOTES TO EUKARYOTES TO YOU

All living things are made of cells, and cells can be divided into two main types: prokaryotic and eukaryotic (see pages 48–9). The prokaryotic cells came first and ruled the world. They are still around, in the archaea and bacteria that flourish in all environments. They are also inside every one of us, and in every other eukaryotic organism. Eukaryotes, including all the multi-celled organisms such as plants and animals, have cells that contain organelles enclosed in membranes. Organelles are tiny cellular organs that carry out particular functions in the cell. And some of those organelles are ancient prokaryotic cells that were captured and enslaved by our early eukaryotic ancestor cells long, long ago – perhaps 2.5 billion years ago.

SYMBIOSIS – A HELPING HAND

Symbiosis is common in nature. It's the mutual assistance that two organisms give to each other, working in a co-operative partnership. You have a symbiotic relationship with your gut flora (the microbes in your gut). Without certain bacteria in your gut, you wouldn't be able to digest your food. And those bacteria you rely on, rely on you, too, having evolved to live inside you and gain all their own nourishment from the food you eat. At a more visible level, flowering plants rely on the insects that in turn depend on them: the insects pollinate the plants and in return feed on them. Large animals such as rhinos and hippos benefit from birds removing parasites, while the birds benefit from a meal of parasites. Often, species co-evolve; some flowers are pollinated only by one type of insect, for instance. If one of the pair is threatened with extinction, so will their partner be threatened.

SYMBIOSIS IN CELLS

Scientists first noticed that mitochondria inside cells look very like bacteria. In the 1970s, Lynn Margulis gave compelling evidence that early proto-eukaryotic cells engulfed, but did not consume, prokaryotic cells and entered into a symbiotic relationship with them. Together, living in endosymbiotic harmony, they produce eukaryotic cells. The prokaryotes had a place to live, with a supply of the organic and inorganic molecules they needed, and the eukaryotes outsourced the extraction of energy to them.

Here's how it might have happened:

1. First, a prokaryotic cell lost its cell wall. The flexible membrane beneath the cell wall began to grow and fold in on itself. This produced membrane-bounded areas within the cell, eventually becoming a nucleus (which held the cell's genetic material) and other structures. By this means, it became a primitive eukaryotic cell.
2. The primitive eukaryote developed so that instead of just ingesting useful molecules from the water around it, it became able to ingest smaller prokaryotes.
3. At some point, the eukaryote took in an autotrophic prokaryote but didn't digest it. Autotrophic organisms can produce their own food; the prokaryote in this case was probably able to photosynthesize in some way. This turned out to benefit both, as the prokaryote was protected and had a supply of nutrients, and the eukaryote gained energy from the prokaryote's cellular respiration.
4. Another type of absorbed prokaryote eventually became the mitochondrion that is found in the cells of most eukaryotes, including animals, plants and fungi. As the relationship crystallized into something permanent, the absorbed prokaryote lost some of its genes, and kept just those that were needed.
5. Plant cells took a further step, engulfing a photosynthetic cyanobacterium that eventually became the plant organelle that carries out photosynthesis, the chloroplast.

It's possible that further organelles in eukaryotic cells also had their origins in prokaryotes that had been engulfed but not digested.

WHAT ARE THEY DOING IN THERE?

Mitochondria and chloroplasts are both concerned with producing and storing energy for cells. Energy is stored in the form of chemical bonds. Mitochondria store energy in a chemical called ATP (adenosine triphosphate), which has

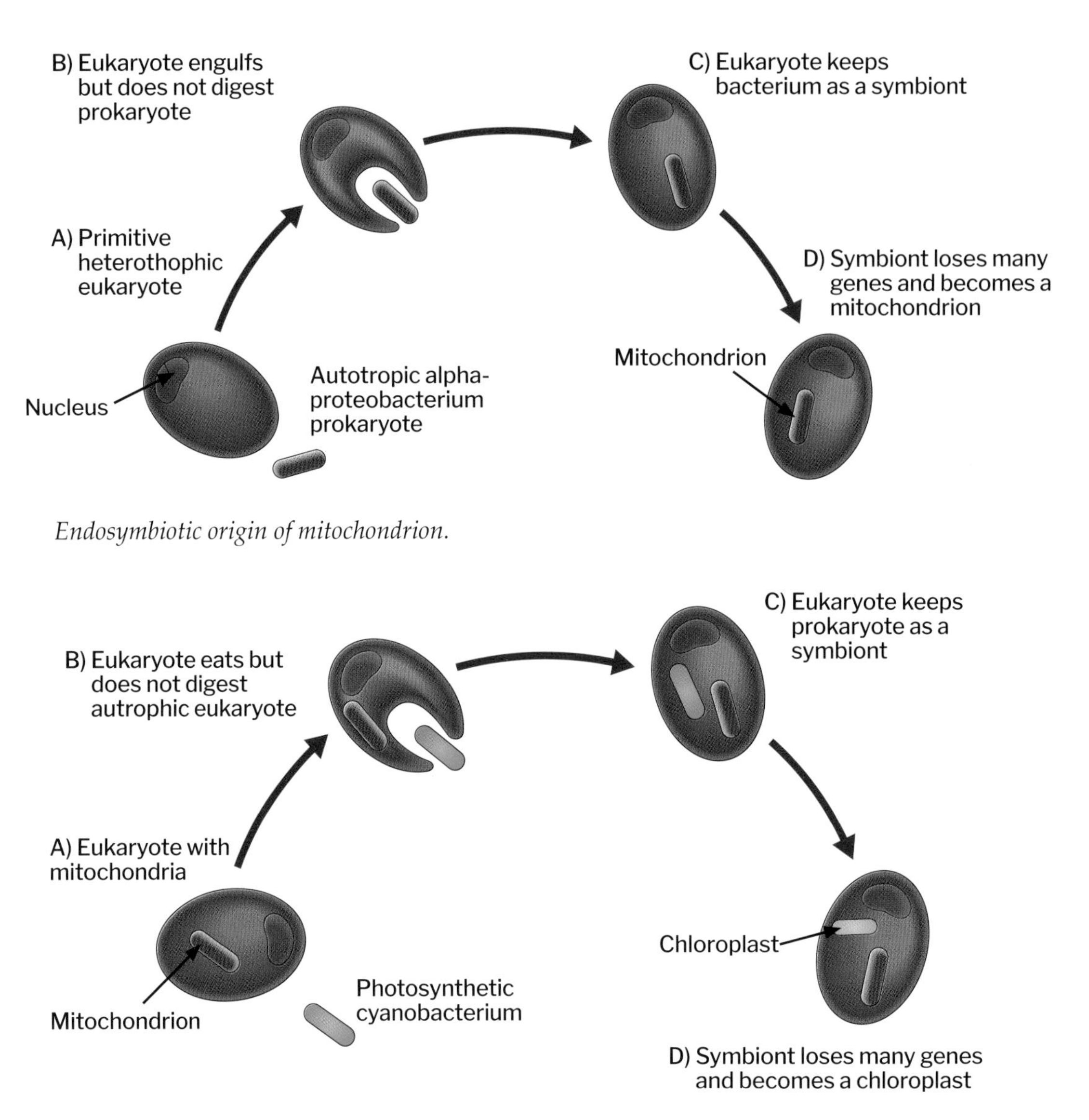

Endosymbiotic origin of mitochondrion.

Origins of mitochondria and chloroplast.

high-energy phosphate bonds. When a cell needs energy for its biochemical functioning, ATP can be broken down to release energy, leaving ADP (adenosine diphosphate). The ATP is later reformed: it's rather like having a rechargeable battery. Mitochondria are also involved in cell growth and reproduction.

A cell usually has many mitochondria (some more than others, depending on their energy needs). These are joined together in mitochondrial networks, which are dynamic and change frequently. Mitochondria reproduce within the cell and have their own (reduced) DNA. In sexually reproducing organisms, mitochondrial DNA is inherited from the female parent as egg cells contain mitochondria, but sperm cells do not.

Green plants have chloroplasts as well as mitochondria. The chloroplasts carry out photosynthesis, creating glucose and oxygen from carbon dioxide and water using the energy of sunlight to drive the reaction. Glucose is the plant's 'fuel'; it's used to grow, form flowers and seeds, and develop fruit. Glucose molecules can be combined and configured (sometimes with the addition of other chemicals) to produce cellulose, carbohydrates and amino acids (the building blocks of proteins). These are the larger molecules that make up the plant's body. As animals can't produce glucose to synthesize these vital chemicals, they rely on eating plants – or eating other animals that have eaten plants – to gain a source of energy and nutrients. Even the most dedicated meat-eater relies ultimately on plants, and on the prokaryotes that became their chloroplasts, to sustain life. If endosymbiosis had never happened, we would have none of the large eukaryotic organisms that now throng the planet.

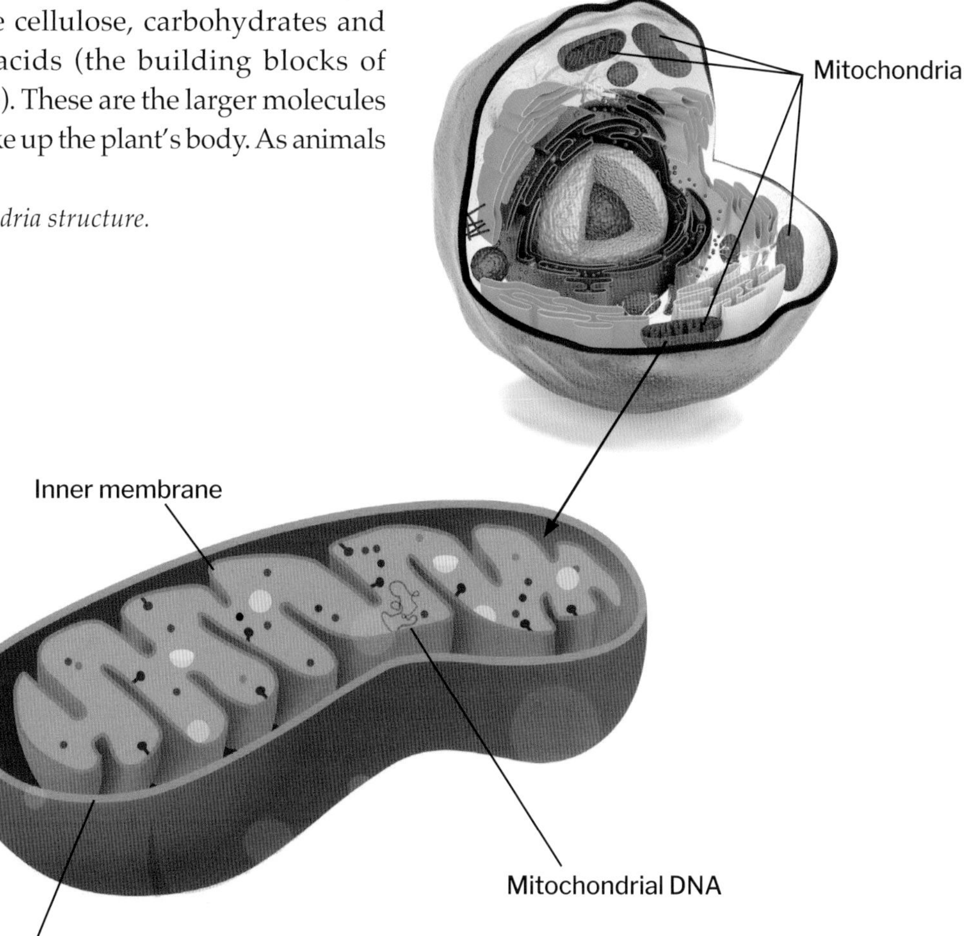

Mitochondria structure.

36

Vax v. anti-vax

Vaccination

For most people, getting their childhood vaccinations, and any they need for travelling, has been non-contentious. But over the Covid years, vaccination became a hot topic. The pressure to vaccinate most of the world during the pandemic led to more debate and abuse than at any time, perhaps, since vaccination began. The science of vaccination is simple, but the social aspects have become complex.

VACCINATION AND COWS

The word 'vaccine' comes from the Latin *vacca*, meaning cow. It was so named because the first vaccine, to protect against smallpox, was made from matter from cowpox pustules. Cowpox is a disease related to smallpox, but much milder. It causes ugly boils on the hands of people who milk cows by hand, but smallpox is deadly. Smallpox causes a body-wide rash and extreme fever. It can lead to blindness, brain damage or death (it has 30 per cent mortality) and always leaves disfiguring scars. For millennia, smallpox ravaged the world, until it became the first disease to be wiped out by vaccination. The vaccine was developed and widely used by English physician Edward Jenner, who vaccinated his first subject in 1796. He was not the first to

Edward Jenner.

A nasty blow

Centuries ago, people in China, India and parts of Africa practised 'variolation', which involved inoculating healthy people against smallpox by introducing a small dose of smallpox virus to them. This was done either by blowing powdered smallpox scabs into the nose, or putting a little matter from smallpox pustules or scabs under the skin. It usually produced a mild case of the illness from which the patient recovered, but it was not without risk: 1–2 per cent of people died, and the disease could be passed on to others. It was introduced to Europe by Lady Montagu Wortley in 1721. The same year, an enslaved man known as Onesimus, described the procedure to the minister and writer, Cotton Mather, and it began to be used in America.

notice that cowpox bestowed immunity to smallpox, nor the first to use cowpox to induce immunity, but he pioneered its widespread use.

TRICKING THE BODY

Jenner did not know how his vaccine worked in the body; smallpox is a viral disease, and viruses were not even identified until 100 years later. But he didn't need to know how it worked in order for it to be effective. We now know that a vaccine works by preparing the body's immune system to fight a particular pathogen.

THE BODY FIGHTS BACK

When the body is attacked by a pathogen, such as a bacterium or virus that causes disease, it mounts an assault on the unrecognized invader. It can do this because pathogens typically have chemicals on their surfaces which don't match those usually found in the body. The first line of defence is the innate immune system. Immune cells called phagocytes and natural killer cells attack, destroy and remove bacteria and infected cells. This can spring into action immediately. The second line of defence is the adaptive immune system. This makes antibodies to counter specific antigens. It stores information about antigens and the antibodies made and can make the antibodies again quickly if needed. This means that if you have a disease such as chickenpox, your body makes antibodies to fight the chickenpox virus, and if you are exposed to the disease again later you are unlikely to get ill with chickenpox as your immune system can immediately produce the right antibodies to prevent it taking hold.

Vaccines work by showing your body the antigen, prompting it to make antibodies to counter it. Although you don't have the disease, it remembers the antibodies and if you are later exposed to the same antigen, it can launch a rapid

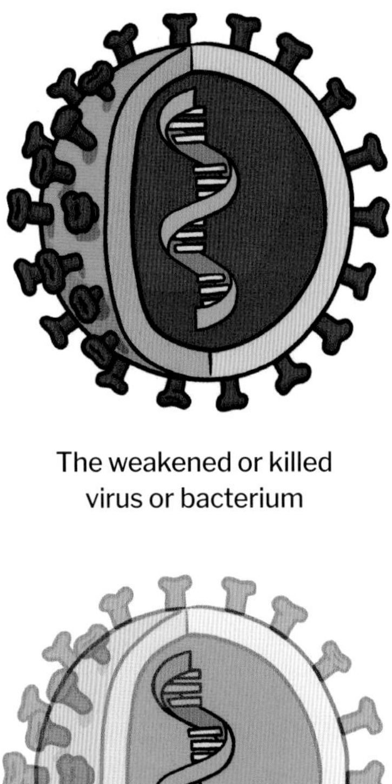

The weakened or killed virus or bacterium

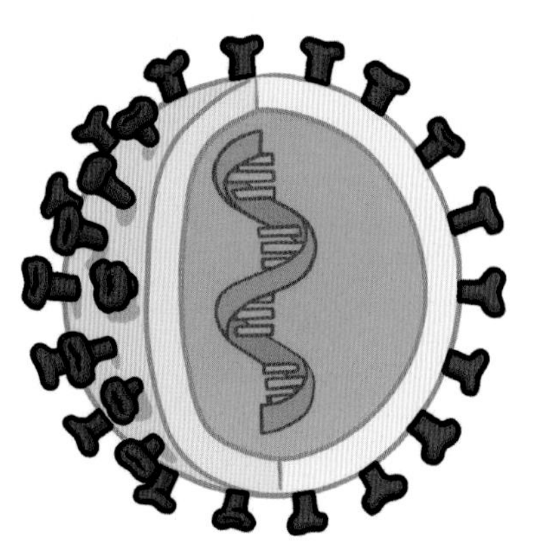

Parts that trigger the immune system

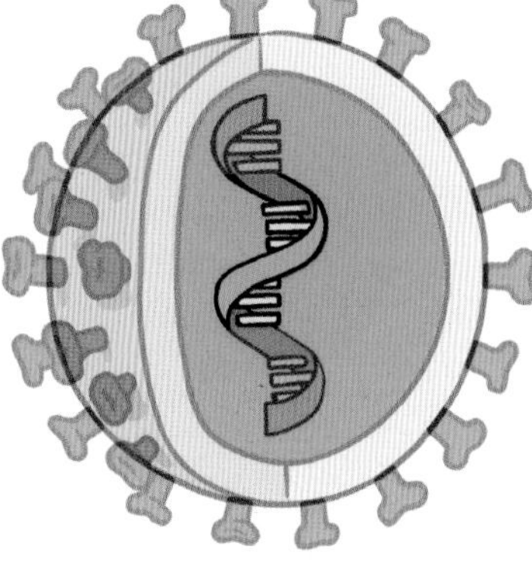

A tiny fragment of mRNA.

Three types of vaccine.

response to protect you. The stuff of the vaccine doesn't spend long in your body – it's soon dismantled. But the 'recipe' for the antibodies that are created is stored for a long time.

ATTENUATED, INACTIVATED AND MRNA VACCINES

There are three main types of vaccine in widespread use. An attenuated (or 'attenuated live') vaccine contains a small quantity of the pathogen that causes a disease. It has been treated so that the pathogen is weakened and shouldn't cause disease. However, people with a compromised immune system, or caring for people with a poor immune response, should take medical advice before using live attenuated vaccines. Attenuated vaccines produce the strongest immune responses, and often only one or two doses provide lifelong protection. Vaccines for measles and rubella use attenuated virus.

An inactivated vaccine contains not just weakened but killed germs. The immune response is not as strong. Inactivated vaccines are used against rabies, polio and flu.

An mRNA vaccine doesn't use any form of the original pathogen. The only mRNA vaccine currently in use (as of 2022) is against Covid-19. The vaccine contains just a fragment of mRNA, or messenger RNA, which plays a role in

Eighteenth-century cartoonists had fun lampooning Jenner's vaccine based on cowpox.

copying DNA and proteins. The mRNA fragment contains the instructions for making the proteins present on the spikes on the surface of the virus. It is this protein that the immune system uses to recognize the virus. A few cells in your body follow the instructions in the mRNA to make the spike protein, which the immune system then immediately recognizes as alien. Your body builds antibodies against the disease just from exposure to the protein, with no risk of infection. The mRNA fragment breaks downs in a few hours and the proteins created are soon destroyed by the body. The immune system 'remembers' how to make the antibodies it created to destroy the protein, giving future protection against the disease.

VAX AND ANTI-VAX

When Jenner introduced his first vaccine, many people ridiculed it and him. Cartoons showed people sprouting cow-like appendages. There has been an anti-vax movement ever since, sometimes more powerful than at others. Over the last two centuries, vaccines have saved millions of lives. Undoubtedly, there have been some errors with vaccine manufacture and use, as with any product. But vaccination has also been vilified, on the basis of flawed and even deliberately falsified research, and on the basis of misinformation and bizarre conspiracy theories. During the Covid-19 pandemic, some vaccine-hesitant people worried about side-effects. Others believed nonsensical suggestions that the vaccines gave control of their minds to someone else, or would kill large portions of the population, or even turn people into crocodiles.

37

Holes in space?

Black holes

The popular conception of a black hole is as a hole in space that things get sucked into if they stray too close. In fact, a black hole is entirely the opposite of a hole. A hole is where there is a gap in matter. A hole in your pocket is where there is no pocket-fabric. But a black hole is somewhere that has so much matter it creates an incredibly strong gravitational attraction that drags in everything – and so it becomes even more dense.

THE DAWN OF BLACK HOLES

The first hint of a black hole in the history of astronomy was the English astronomer John Michell writing in 1783 of objects so dense their light can't escape. If no light can come from an object, it will naturally appear black. Albert Einstein's general theory of relativity (see page 186) predicted the existence of black holes, suggesting that as gravity is the distortion of spacetime by mass, a very great concentration of mass could distort spacetime enough to produce the extreme gravity Michell pondered, creating an area from which light can't escape. The term 'black hole' was first used in 1964.

MAKING HOLES

American scientists J. Robert Oppenheimer and Hartland Snyder suggested in 1939 how a black hole might come about. They proposed that when a massive star collapses under the weight of its own gravity at the end of its life, it will become increasingly dense. As more of the star falls in towards the middle, it becomes denser and denser. Its gravitational attraction for matter increases further, and matter is drawn further inwards. Eventually the matter of the star would be packed into such a small volume that it would be so dense its gravity would not allow even light to escape. It would become a black hole. An 'event horizon' around the black hole marks the boundary which matter and energy can't cross without being drawn in. This is now accepted as the way many black holes form.

When stars more massive than the Sun run out of hydrogen to fuse into helium, they fuse heavier elements in turn until there is a core of iron

Two black holes collide and merge (false colour).

remaining. While they continue with nuclear fusion, pressure from the energy created at the heart of the star streaming outwards counteracts gravity drawing the star inwards. But when the star can no longer continue with fusion, gravity is not counterbalanced by any other force. It then drags all layers of the star inwards. There is no room at the centre for all the mass, though. It bounces off in a massive explosion, called a supernova. While much matter is hurled outwards into space, the super-dense middle of the star remains, becoming a neutron star or a black hole.

MASSIVE HOLES

Black holes as the residue of dead stars account for regular-sized black holes with the density of large stars. But in 1974, astronomers Bruce Balick and Robert Brown discovered a supermassive black hole at the middle of the Milky Way. Called Sagittarius A*, it has a mass four million times that of the Sun. It turns out that there are supermassive black holes at the centre of galaxies throughout the universe. Indeed, the galaxy M87 contains a black hole billions of times the mass of the Sun. No one is sure how supermassive black holes form. One suggestion is that a star about 1,000 times the mass of the Sun becomes a regular, if rather large, black hole and then grows slowly as it draws in surrounding gas and merges with other black holes. An alternative is that it starts very large: a huge cloud of gas collapses to create a

A disc of matter whirls around a black hole, and jets of high energy particles are thrown outwards.

supermassive star, perhaps a billion times the mass of the Sun. This would be very fragile and last only a few million years before collapsing in on itself to make a supermassive black hole.

EATING EACH OTHER

Black holes can grow not only by drawing in matter that crosses their event horizon, but by merging with other black holes. Gravitational waves believed to come from two black holes colliding were discovered in 2016.

LOOKING FOR DARKNESS

We can see stars in the sky because they give out light and other energy. If no light or other energy can escape from a black hole, that clearly makes them very difficult to see. Astronomers can often find a black hole by noticing how objects nearby behave. The gravity of a black hole

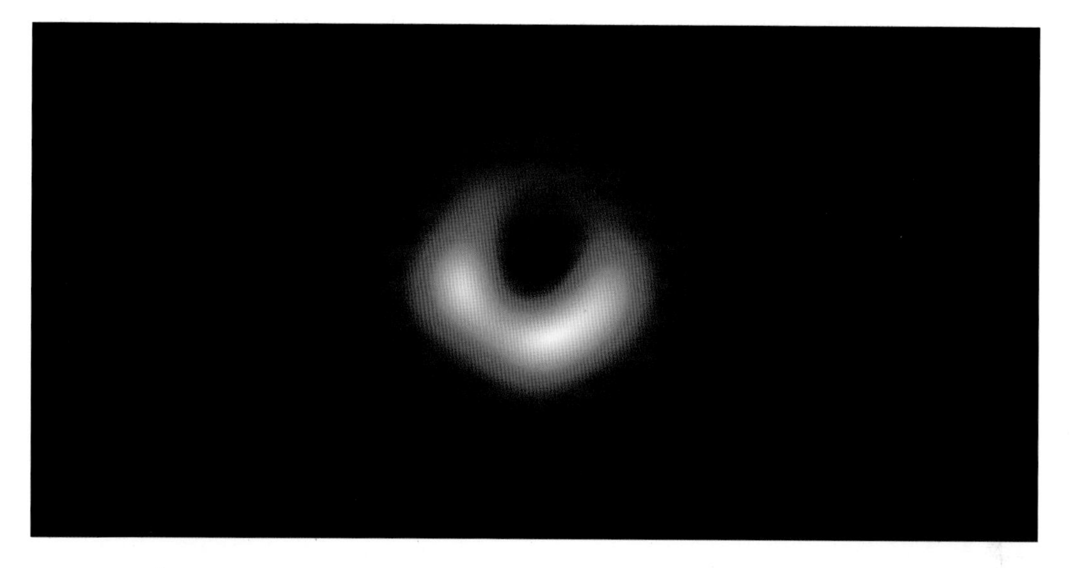

Photo of a black hole taken by Event Horizon Telescope, 2019.

can cause things to swirl around a central point which might look empty but is actually occupied by a very massive but invisible black hole. As matter is drawn towards a black hole, it whirls around it, often becoming very hot. We can detect this heat with telescopes. As matter is pulled into the black hole, some energy is released as light which bounces out, as it's not crossed the event horizon. We can sometimes detect this, too. Telescopes such as the Chandra X-ray telescope can identify the signs of matter entering a black hole.

The first photo of a black hole was taken in 2019 by the Event Horizon Telescope, used specifically to investigate black holes.

We're safe here

People sometimes worry that the Sun could become a black hole and Earth would be drawn into it. The Sun is too small to become a black hole. When it eventually dies, in around 4.5 billion years, it will first expand massively, and Earth will be doomed at that point. But it won't ever collapse into a black hole as it doesn't have enough mass. Even if it were a black hole, it would have the same mass, and so the same gravity, as it has now. We are not drawn into the Sun, so we wouldn't be drawn into it if it were a black hole with the same gravitational impact.

38

Round and round

Biogeochemical cycles

Earth is the ultimate recycler. All the important chemicals that organisms need are constantly recycled. The oxygen you breathe today was once breathed by dinosaurs; the carbon in your cells was once in the rock of the seabed. Although we, as humans, might be exceptionally bad at sustainable living, the planet is an expert.

Earth is a closed system; apart from energy from sunlight, very little enters or leaves the planet.

CYCLES AND SPHERES

The chemicals on Earth can exist in any of four places: the atmosphere (the gassy envelope around the planet); the biosphere (the realm of living things); the hydrosphere (Earth's water); and the lithosphere (rocks). Chemicals move between these as they're recycled. In general, cycling through the atmosphere, biosphere and hydrosphere can be very rapid, but movement through the lithosphere is often very slow.

SEEING CYCLES

Some chemical cycles have been known for quite a long time. In its most basic form, the water cycle is fairly obvious. Water evaporates from the surface of the sea and land, and is released by organisms, and then collects in the atmosphere as a gas, becoming clouds when it forms droplets that clump together. When a cloud is saturated, the water falls as rain. Some seeps into the ground, some runs over it, some is consumed by plants or animals, and some evaporates directly from where it falls. Surface water collects in rivers which return to the sea. The water begins its cycle again.

There are equivalent cycles for other chemicals, including nitrogen, oxygen, carbon, sulphur and phosphorous. Nitrogen is 'fixed' from the air by certain types of bacteria, then taken up by plants and built into their cells. The plants are eaten by animals, or die and are dismantled by decomposers. The animals might be eaten themselves, but eventually the nitrogen they take from plants either comes out in their waste material or stays in their bodies until they die. Then that, too, along with their waste and the dead plants, is broken down by decomposers and the nitrogen is returned

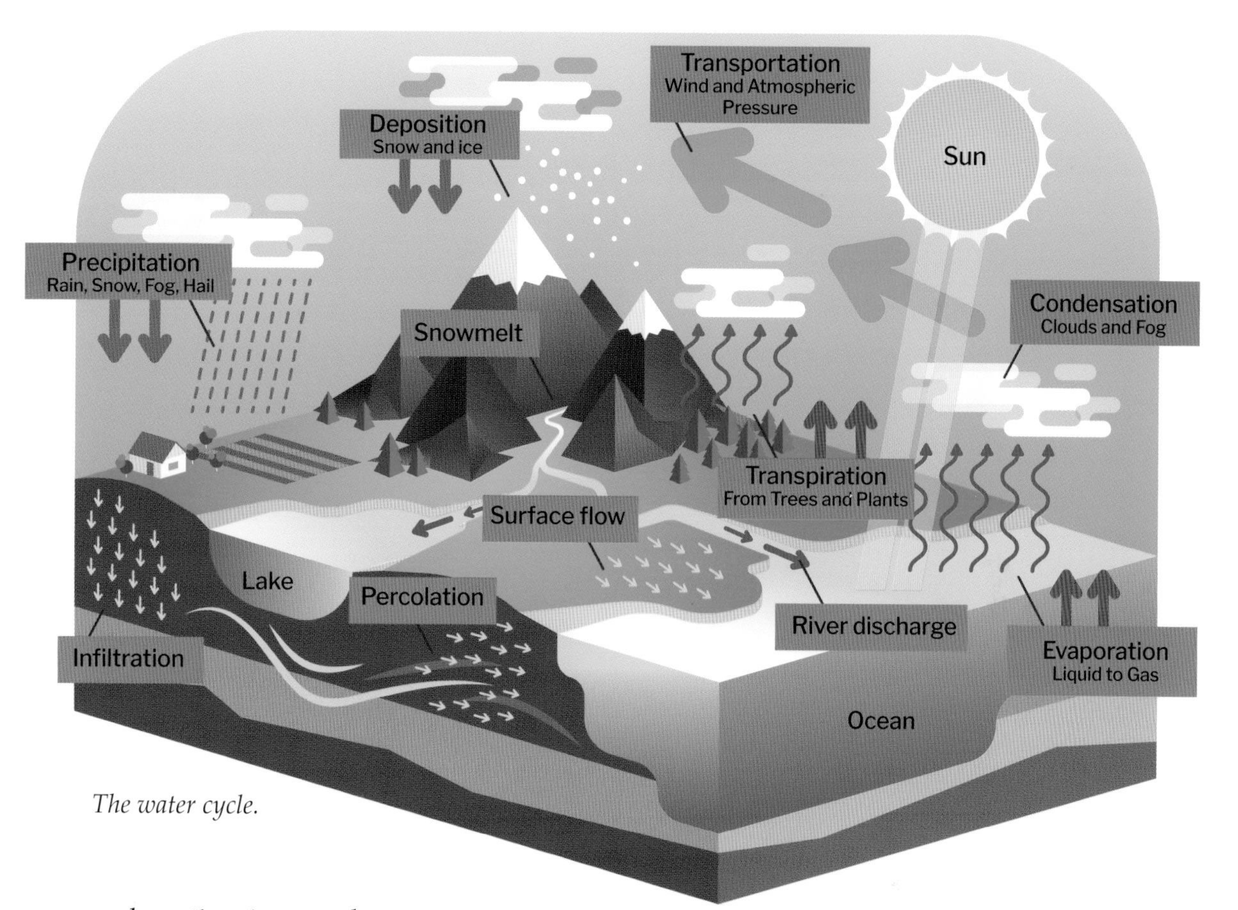

The water cycle.

to the soil and atmosphere. Round it goes again. Today, we also make nitrogenous fertilizers which short-circuit some of this, taking nitrogen from the atmosphere to make fertilizers which we put directly into the soil to help plants grow.

MESSING WITH CARBON CYCLES

Carbon runs through two cycles in parallel. One is biological and is sometimes called the fast carbon cycle. The other is geological and is called the slow carbon cycle. In the biological carbon cycle, plants take carbon dioxide from the atmosphere and, with water from the ground and energy from sunlight, produce a sugar (glucose) and oxygen. They release the oxygen into the atmosphere. This is called photosynthesis and it is essential to most life on Earth. The oxygen is used by respiring organisms, including plants and animals. The glucose is used by the plant to build its cells and run its metabolism. The cells of the plant provide food for plant-eating animals, and they in turn provide food for meat-eating animals. When plant and animal waste decomposes, the carbon is released again. Sometimes, in certain circumstances, this waste doesn't decompose but fossilizes over thousands

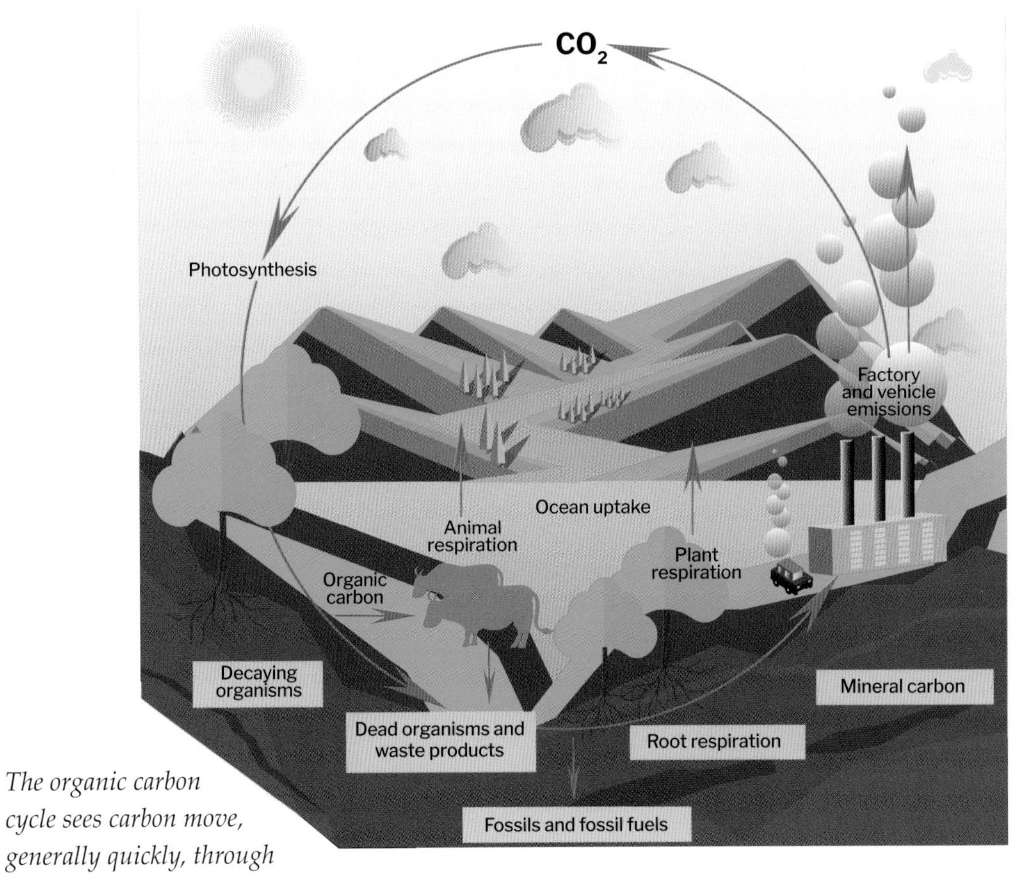

The organic carbon cycle sees carbon move, generally quickly, through living organisms and the atmosphere.

of years. This produces fossil fuel, such as coal or oil. The carbon is held suspended in that fuel indefinitely.

At the same time, a much slower cycle is going on. Carbon is fixed with oxygen in carbonate rocks. Over very long periods of time, these rocks weather – they are worn away by acidic rain that forms as carbon dioxide in the atmosphere dissolves in water to make carbonic acid. Dissolved ions from the rocks are carried to the ocean where they combine with bicarbonate ions. Most of this is calcium carbonate, and most of the calcium carbonate is first used by sea creatures building their shells from it. When they die, their shells fall to the seabed and are eventually compressed into rock. Tectonic movement (see pages 110–13) slowly drags the seabed towards the coast, where it is pulled under the continental crust and down into the mantle, where it melts. The carbon that had been built into the rock is freed and eventually comes out of volcanoes as carbon dioxide, returned to the atmosphere. The slow carbon cycle takes 100–200 million years, so it's not a quick fix for our climate problems.

Where humans have messed up the

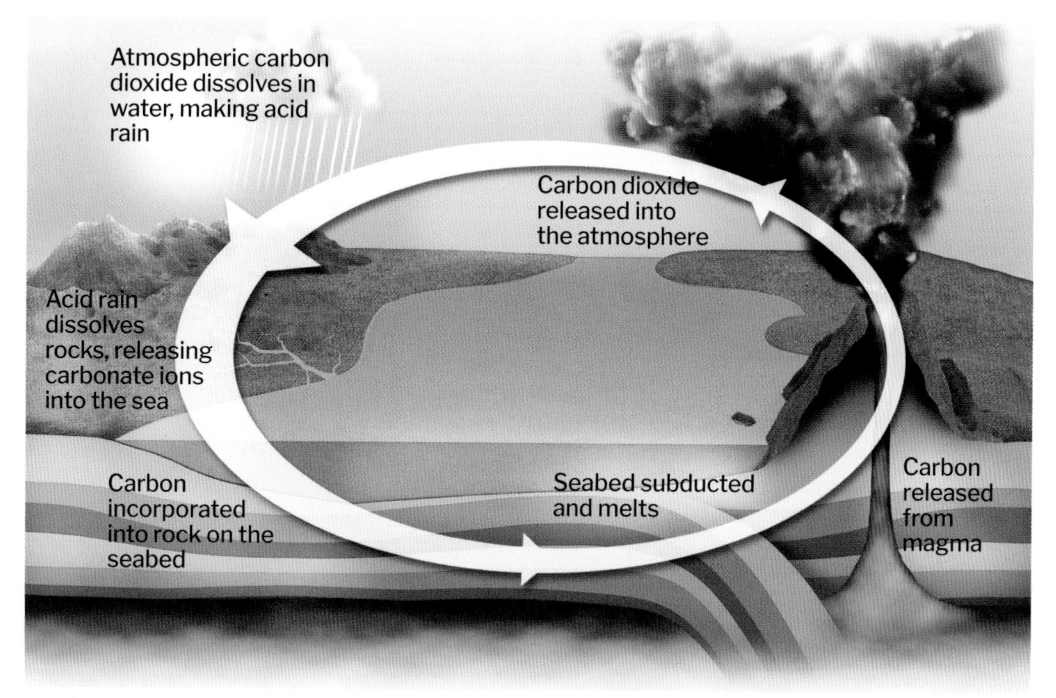

It takes millions of years for a carbon atom to move through the slow carbon cycle.

carbon cycle is by taking fossil fuels, which contain carbon locked away millions of years ago, and burning it in large quantities. This immediately releases carbon that has been out of circulation for a very long time – up to 350 million years. It's too much for the fast carbon cycle to deal with and we can't wait for the slow carbon cycle to take care of it.

Cycles within

Living bodies, too, have their own biochemical cycles. One of the most important is the cycling of ATP (adenosine triphosphate) which acts as a kind of energy bank within cells. Chemical reactions either use or produce energy. The energy is involved in making and breaking the bonds between atoms to form and reform molecules. A living body relies on reactions of both types, some that need energy and some that release energy. A cell balances the energy needs of reactions using ATP. To carry out a reaction that needs energy, ATP combines with water to produce ADP, which releases energy. When a reaction that releases energy is carried out in the cell, the spare energy is used to change ADP back to ATP and water.

39

A helping hand for the chemistry of life

Enzymes

Enzymes are chemicals which speed up chemical reactions in living organisms. They are catalysts, as are all chemicals that speed up reactions or encourage them to take place. Catalysts are not permanently changed in a chemical reaction; the same quantity of catalyst is left at the end as was present at the start. Unlike many other catalysts, enzymes are very specific in terms of which reactions they affect.

ENZYMES EMERGE

The first enzyme discovered was diastase, found in 1833, though the term 'enzyme' wasn't used until 1877. Diastase catalyses the breakdown of starch into maltose (a sugar). Nearly all the reactions that take place in cells as part of life processes rely on enzymes. Enzymes catalyse more than 5,000 different types of reaction, and life as we know it would be impossible without them.

HOW ENZYMES WORK

Enzymes work by reducing the amount of energy that a reaction needs in order to take place. Each enzyme has an 'active site' which is shaped to hold only the specific molecules the enzyme acts on, called the 'substrate'. By holding the reacting molecules in close contact, it enables the reaction to take place quickly. Without the enzyme, a reaction will only take place when molecules of the substrates collide during their random movement. The active site has two parts. The catalytic site is where the reaction takes place; next to it are one or more binding sites, which turn the substrate molecule around as necessary to orient it suitably for the reaction.

When an enzyme is present, the reaction takes place when the necessary substrate molecules collide with the (much larger) enzyme molecules and are drawn into the active site. Increasing either the amount of substrate or the amount of enzyme increases the frequency of collisions and so speeds up the reaction – but only to a certain point. After an optimum concentration, adding more has no effect.

Fast and furious

The effect of enzymes is dramatic: some speed up a reaction to millions of times the rate it would take place without the enzyme. One of the most striking is OMP decarboxylase, which facilitates the synthesis of pyramidine (a component of DNA). Without the enzyme, the half-life for this reaction is 78 million years; with it the half-life is 18 milliseconds, an increase in reaction rate by around x10^{17}.

LOCKS AND KEYS OR BENT OUT OF SHAPE?

There are two models for how enzymes work. One, the 'lock and key' model, has the enzyme's active site exactly the right shape for the substrate molecules. Only one type of molecule will fit, just as only the right key will undo a lock.

The other model, the induced fit hypothesis, suggests a less perfect fit between enzyme and substrate, but when the substrate binds to the active site the enzyme changes shape to fit to it precisely.

Mechanism of enzyme–substrate interaction

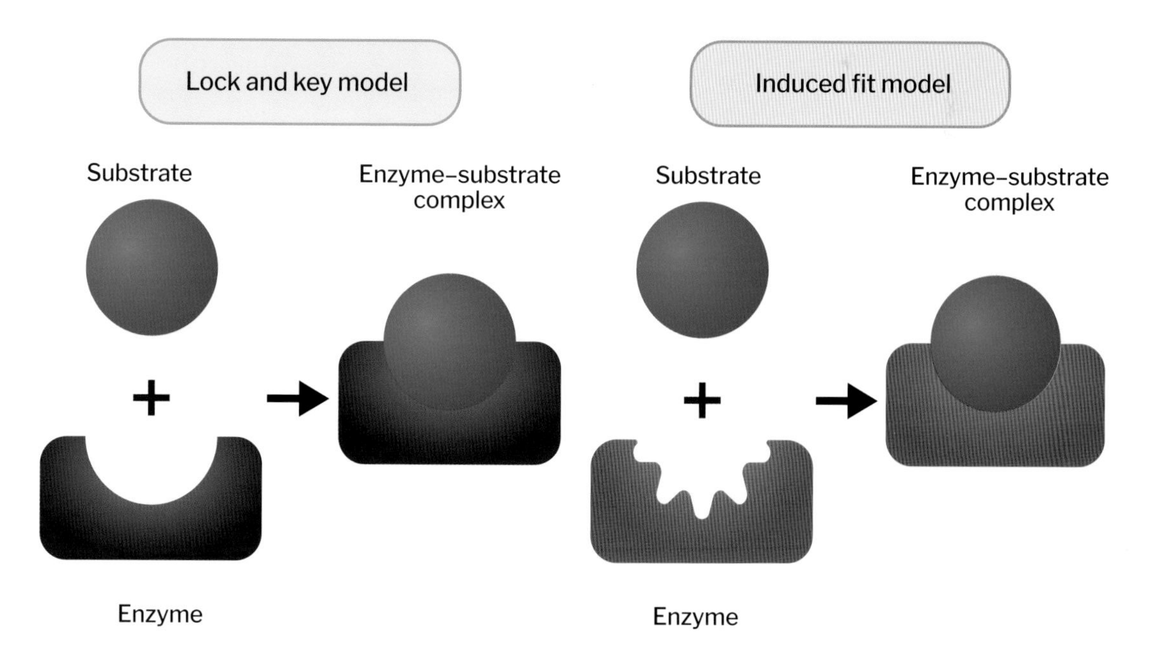

The 'lock and key' and 'induced fit' models of enzyme action.

EASILY UPSET

Enzymes are very sensitive to environmental conditions and won't work outside a narrow temperature and pH range. As proteins, they can be denatured – permanently bent out of shape – by high temperatures or conditions that are too acidic or alkaline.

Enzyme action can also be disrupted by 'enzyme inhibitors'. These are chemicals that block the active site's ability to bond with the substrate so that the enzyme then can't do its job. They can work in either of two ways. A 'competitive inhibitor' mimics the substrate and blocks the site so the substrate can't get to it. As they aren't able to take part in the reaction, the molecules of the inhibitor just sit there making the enzyme inactive. A 'non-competitive inhibitor' binds to some other part of the enzyme, not the active site. In doing so, it bends the enzyme out of shape. The active site is then distorted and won't fit the shape of the substrate. Some enzyme inhibitors act as poisons, preventing bodies functioning in some way. Others are medicines, as they can block destructive processes that are catalysed by enzymes.

LIVING OR NOT?

Initially, there was confusion between enzymes and fermentation. When Louis Pasteur discovered the role of yeast in the fermentation of sugar to produce alcohol, he accounted for it by referring to a 'vital force' in the yeast cells. In 1897, Eduard Buchner discovered that the yeast cells didn't need to be alive for fermentation to occur. He concluded an enzyme in the yeast, which he named zymase, is responsible for fermentation. The word 'enzyme' is now used for a non-living chemical catalyst and 'fermentation' is chemical activity produced by living organisms.

40

Layer after layer

Superposition

While the rocks of the seabed are relatively new, being never more than about 250 million years old, those of the landmasses can be billions of years old. We can read the story of how our lands have been built by looking at the strata (layers) of rock that make up the land.

MAKING ROCK

Rocks on Earth are divided into three types: igneous, sedimentary and metamorphic. Igneous rock originates inside Earth, as magma. It either emerges through volcanoes and at rift zones and hardens on the surface, or hardens within the crust. Typical igneous rocks include granite and basalt. Sedimentary rock is made from hardening sediment. Sediment consists of tiny bits of sand and organic matter from living organisms. As more sediment collects above it, heat and pressure fuse the lower deposits into rock. Shale and slate are examples of sedimentary rock. Metamorphic rock is formed when either igneous or sedimentary rock is heated enough to change chemically, but not to melt entirely. Marble is a familiar metamorphic rock.

LAID IN LAYERS

Throughout Earth's long history, rock has been laid down in layers. Igneous rock has been overlaid by sediment which has turned to sedimentary rock. Much of the seabed is igneous rock, formed at the sites of seafloor spreading. Because it moves slowly to the subduction zones where it is drawn back into the mantle, the rock of the seabed is constantly recycled, and so not very old. On land, the story is rather different. Volcanic eruptions have added layers of further igneous rock, and layers of volcanic ash, over sedimentary rock. Further sediment has been laid down on top. If Earth's surface were peaceful, it would be easy to read history in the layers of rock, undisturbed for billions of years, lying in the order in which they were laid down.

The Danish scientist Nicolas Steno (1638–86) was the first person to describe superposition – the laying down of rock in layers, with the oldest at the bottom and the most recent at the top. He recognized that undisturbed layers allow relative dating of rocks. Steno pointed

Undisturbed rock strata.

out that rock is laid down in horizontal layers, and that if anything cuts across a layer it must have come later, after the rock had been deposited.

The movement of tectonic plates (see page 112), and the heat and pressure it produces, has changed some of the rock to metamorphic forms and has disrupted the layers. The result is, in some places, a jumble of layers that have been uplifted, dropped, twisted, partly melted and otherwise disrupted, making strata that rise in rucks and wrinkles or even lie vertically.

LIFE AND DEATH IN THE ROCKS

Some sedimentary rocks contain fossils. Occasionally, an organism that dies is covered quickly with sediment before it is broken up or eaten. It might then undergo chemical changes as heat and pressure form sedimentary rock, trapping relics of the organism within. It's clear from Steno's rule of superposition that a fossil in a lower layer of rock is older than one in a higher layer of rock if the layers are undisturbed. This became the key to unlocking the dating of layers.

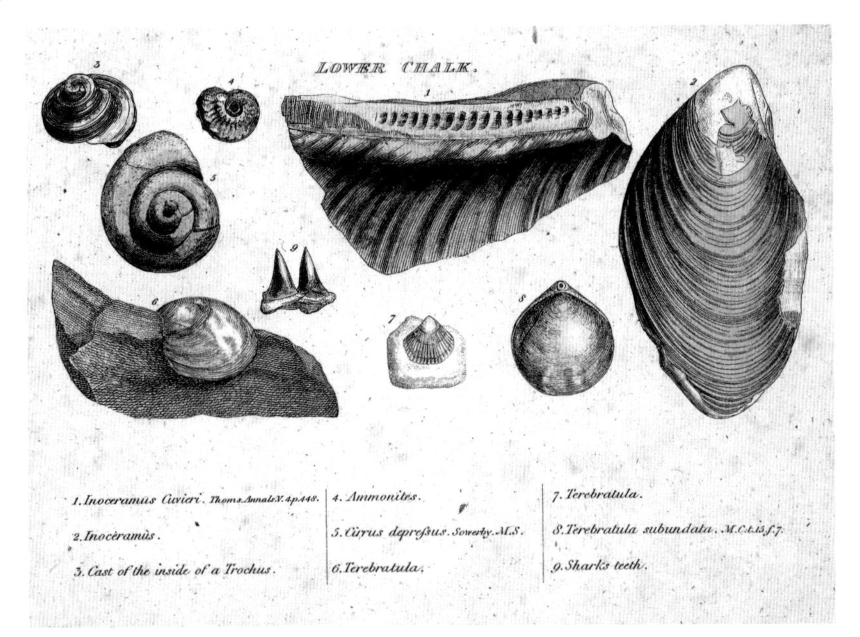

William Smith's guide to fossils.

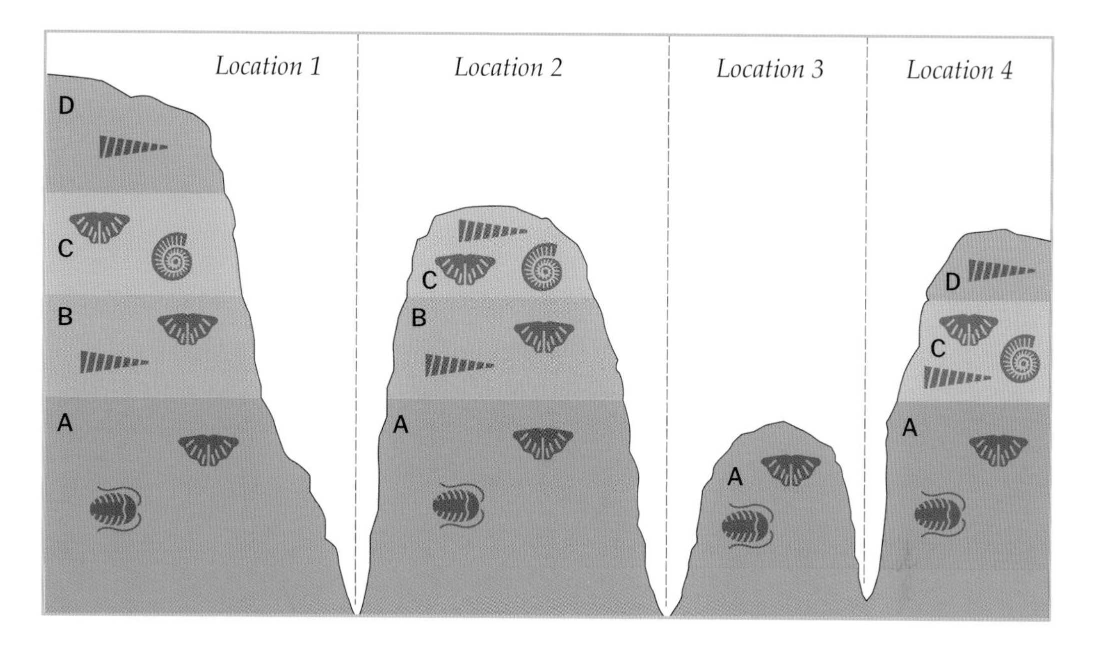

Index fossils in strata.

The British canal engineer William Smith mapped the rock strata he found in his work, noticing that similar sequences of strata occurred in different places. He also found similar strata contained similar fossils, and fossils change between strata. He realized he could compare fossils to decide whether a rock stratum in one place was the same as a stratum he found elsewhere, so introducing the method of relative dating of rocks by reference to index fossils. He labelled as 'faunal succession' the changing nature of the fossils. This early clue that animals change over time would later become evidence for evolutionists.

Today, rock layers are still dated by the presence of index fossils, but this alone gives only relative dates: one layer is older or younger than another. Now, radiometric dating is used to gain absolute dates (a specific number of years) for rocks. When igneous rocks form, they contain some radioactive elements. These continue to decay (see page 60). By measuring the amount of the original radioactive element and the amount of 'daughter' elements produced by its decay, it's possible to work out the age of the rock. Fossils, though, occur mostly in sedimentary rock. Geologists look for layers of igneous rock that 'bracket' layers of sedimentary rock so that they can provide an age range for the sedimentary layer. Radiometric dating can be used directly to date some very young fossils (up to 50,000 years old), but in fossils the element measured is carbon, and the half-life of carbon is too short for carbon-dating to work on older fossils.

41

Never out of energy

Conservation of energy

When you're feeling tired at the end of the day or after a bout of exercise, the idea that energy is conserved might seem unlikely. If it's conserved, why do you have no energy? Even if energy has gone out of you, it hasn't just gone. The total amount of energy still exists, it's just in different places or different forms. The concept of energy was introduced in the 1740s by the French physicist Emilie du Châtelet (1706–49), and she also set out the principle of the conservation of energy.

PASS IT ON

In what physicists refer to as a 'closed system', energy is just passed around. It might change form – from movement (kinetic energy) to heat, for example – but the total energy of the system doesn't change. If you sit in a sledge at the top of a snowy slope, you have potential energy. You've gained that potential energy by moving a mass (yourself and the sledge) a distance upwards against gravity. The amount of gravitational potential energy you have is equal to the mass (you and the sledge) multiplied by the height in metres you have raised yourself, multiplied by the force of gravity (9.8 m/s^2 (32.2 ft/s^2)). So if the mass is 100 kg (220 lb), and you have gone from 40 m (130 ft) above sea level to 50 m (165 ft) above sea level, you have 100 kg x 10 m x 9.8 m/s^2 = 9,800 kg m^2/s^2. Energy is measured in Joules; this is 9,800 Joules, or 9.8 kJ.

If you then slide down the slope on the sledge, the potential energy is converted to kinetic energy, moving you along. Friction causes some of the kinetic energy to be lost as heat energy, melting a little of the snow and warming your sledge. When you and the sledge get to the bottom of the slope, where you started, you will have none of the potential energy left, but the energy will have been used to do some work: moving you, and warming the sledge and snow, and even you, through friction.

PERPETUAL MOTION?

In theory, in a perfect system with no

Emilie du Châtelet.

friction or forces operating to dissipate energy, something set in motion should keep going forever. If an object is not having to overcome any forces to move, it doesn't use extra energy to keep going. This has set people on a search for a perpetual motion machine. But, of course, there will always be friction within a system. Perhaps the closest we have to perpetual motion is spacecraft that are fired off into space, where there is no atmosphere to produce friction with the craft. Even those, though, are subject to the gravity of the Sun and planets.

HEAT DEATH OF THE UNIVERSE?

It might seem as though, if a bit of energy is lost to heat all the time, it will eventually be the case that all the energy has been converted to heat. Some scientists believe this will indeed happen, and that when the energy has spread out so that everywhere is equally warm (or cold) it won't be possible for energy to do useful work. This is called the heat death of the universe. As it's likely to be (at least) hundreds of billions of years in the future, it's not worth worrying about just now.

42

Non-chaotic chaos

Chaos theory

Chaos theory deals with surprise and unpredictability. It's not really chaotic, as there are logical patterns, cause and effect, behind what happens. The problem is that the patterns are so complex, and the variables so many, that we can't reliably predict outcomes. In particular, a very small change in one variable can lead to a very large change in final results.

The classic example of a chaotic system is the weather. Very many factors affect the weather, and weather forecasting can't take account of all the possible variations and their knock-on effects, so it can't be 100 per cent accurate. If we had perfect knowledge, we would be able to predict the weather accurately as there is nothing really random – or 'chaotic' in the everyday sense of the word – about it.

BUTTERFLIES, SEAGULLS AND HURRICANES

Chaos theory is often captured in the image of a butterfly flapping its wings in one place causing a hurricane on the far side of the world a few weeks later. Henri Poincaré first described the sensitive dependence of a system on small changes in initial conditions in 1890, and soon after related it to weather systems and meteorology. It was fully explored by Edward Lorenz in the 1960s, who used a more prosaic seagull as his example, but with the same message: if a seagull flapped its wings at a particular moment, that could change the weather forever. The large impact of a small change in initial conditions is sometimes called the 'butterfly effect' (but never the 'seagull effect').

Lorenz's interest in chaotic systems came about through his work on meteorology. He had run some calculations, and then ran them again to check his result. But the second time, he entered data from a printout that had rounded the figures – though not by much. Whereas his initial calculation had used a figure of 0.506127, he re-ran it using 0.506. The result was a completely different weather scenario. He went on to find equations that described non-linear behaviour and to determine that weather can't be accurately predicted beyond a

Lorenz's plot of a chaotic system in fluid dynamics, showing a characteristic butterfly shape.

period of two weeks. That's how long it takes for undetectable differences in initial conditions to produce very different outcomes.

CHAOTIC LIVES

Other chaotic systems we encounter commonly include traffic and the stock market. Recently important, the spread of a pandemic is a chaotic system. A state of perfect knowledge should make all of these predictable, but we don't have – and are never likely to have – full and perfect knowledge. Mathematic modelling of chaotic systems tries to produce a most-likely scenario by including rather than ignoring the inevitable uncertainty and errors.

CHAOS ISN'T RANDOM

Chaotic systems don't involve random events, but they often end up looking random. Traffic patterns on a busy road will be affected by obvious factors such as weather, but also by tiny factors such as one person waking up late and so deciding to drive rather than cycle, or rising fuel prices putting a few people off driving. An accident will clog up a road, roadworks will slow traffic, and perhaps a news story about air pollution will deter a few drivers. Individual incidents like this can't be fed into a model, but chaotic models accommodate uncertainties.

43

Survival of the fittest

Evolution

The theory of evolution through natural selection was first set out by Charles Darwin in 1859. Although he had no idea how the process could work at a biochemical level, it was clear to him that organisms change over time and that this accounts for the wide variety of plants and animals found in the world.

BREAKING THE CHAIN

Before the beginnings of evolutionary theory in the 18th and 19th centuries, the principal model of the natural world was of a 'chain of being' that put all organisms into a hierarchy in order of their supposed sophistication. Right at the bottom of the hierarchy were inanimate natural objects such as rocks, and right at the top were mythological beings such as angels. This was a structure that reflected the notion in the Abrahamic religions that God had created all life and that humankind was special and separate. Humans were given a duty of stewardship towards the natural world but also the right to use the natural world.

Charles Darwin.

In this chain, humankind was just below the angels, at the top of the natural order. At the bottom were plants, as they can't move around or do much, followed by 'simple' creatures such as sponges and worms. The chain worked up through beings with increasing powers (as they were perceived) so that land-going animals were above fish, and mammals above reptiles. This type of structure was not based on an idea of development over time, as God had supposedly created all the natural world in one go. Even so, the notion of a hierarchy would prove difficult to shake off, and was even reinforced by ideas of steady 'improvement' through evolution.

DARWIN'S GREAT IDEA

Darwin set sail on a round-the-world trip as the geologist and biologist aboard HMS *Beagle* in 1831. Over the course of the voyage, he made notes and collected specimens of animals, plants and rocks which he returned to England. After the journey, he began the slow process of sorting out and writing up his thoughts. He didn't publish until 1859, when it became obvious that Alfred Russell Wallace was going to publish much the same idea. In the end, both men published at the same time, but it's Darwin we think of first. From his observation of similar, but not identical, organisms in different places, Darwin formulated the theory that plants and animals adapt to changes in their environment so that those best suited to conditions (the 'fittest' for those conditions) survive and breed, while those less well suited are less successful and eventually die out. He saw this as a slow and steady process of adaptation.

Darwin's visit to the Galapagos Islands off the coast of Ecuador gave him the chance to see types of animals not found elsewhere, but they all had some similarities to animals elsewhere. The most famous of his examples are the Galapagos finches. These, he deduced, all evolved from a single type of bird that left the mainland at some time in the distant past to inhabit the group of islands. As the islands offered different types of food, finches evolved to eat those foods, with beaks suited to different purposes. Some of the finches ate seeds, others ate insects, and so on. Birds that ate seeds evolved to have short, stout beaks, while some evolved longer beaks for prising insects out of small spaces. Birds with a slightly longer beak would be more successful at rooting out hidden insects. They would therefore gain more food and would be more likely to survive to breed, so passing on their longer beaks to the next generation. In this way, their beaks grew slowly longer, while those of their seed-eating colleagues grew shorter and stronger. He called this process 'evolution through natural selection' as nature selected which organisms survived by means of how successfully they were adapted to their environment. 'Survival of the fittest' didn't mean the most athletic organisms survived, but those that fitted best with prevailing conditions.

NEW FROM OLD

Darwin's theory allows us to go through

Darwin's finches, each with a beak adapted to its feeding habits.

the history of life on Earth with organisms adapting to new or changing conditions, developing new abilities, or changing their bodies to make the most of the environment they find themselves in. The process led to the gradual emergence of increasingly complex organisms but also to increasing diversity. Where there is competition for resources, such as living spaces or food, organisms become increasingly specialized to occupy a niche with less competition. For this reason, habitats such as the rainforest have a vast number of species, many attuned to very specific conditions. Deserts, which have fewer resources, are home to a smaller number of species and many of those are generalists: they can eat a more varied diet and survive a wider range of conditions.

THE EVOLUTION OF HUMANKIND

Darwin knew that his theory would attract criticism from the Church, and this probably contributed to his delay in publishing it. To avoid too much controversy, he didn't explore the

One of the cartoons mocking Darwinism.

implications for humankind in his first book on the subject, though he did tackle it later in *The Descent of Man* (1871). For many people in the 19th century (and even a few now) the notion that humans are no different from other animals and, like all others, have evolved through a series of adaptations, was unacceptable.

Darwin's theory dethroned humans from the special position people felt they enjoyed as God's favoured being, and replaced it with a view of humankind as a super-evolved ape. It also put humankind at the end of a very long line of beings, with the obvious implication that if we evolved from earlier apes, those earlier apes evolved from something even less human-like. Evolution led right back to the 'warm pool' on a primeval Earth where Darwin supposed life probably started with single-celled organisms. This was a big step down from being the pinnacle of Creation, with dominion over all other animals. It was one of the most significant moments in history for the way humans see themselves, and took a long time to be widely accepted.

PROOF FROM UNDERGROUND

When Darwin set out on the *Beagle*, he was already interested in Charles Lyell's ideas about geology (see page 63). He spent a lot of time observing geological formations and thinking about how they indicated a very long age for Earth – another hot topic of debate in the 19th century. As more and more fossils were discovered, they clearly supported the idea that organisms had changed over time. There were fossils of plants and animals sufficiently similar to some still living to be identifiable as probable ancestors. And then there were the fossils of dinosaurs that were emerging in Europe and North America at the time that were evidence of entirely different types of animals – perhaps, it seemed, of types that were not sufficiently successful to survive changes in their environment.

SLOW AND STEADY OR LEAPS AND BOUNDS?

Darwin saw evolution as a very slow process that maintained a steady pace, like Lyell's concept of slow geological change. This clearly demanded a great age for Earth to achieve the diversity we see now. In the 20th century, evolution has come to be seen as less steady but, in the words of Stephen Jay Gould, a 'punctuated equilibrium'. Organisms might remain much the same for a very long time – even millions of years – if they are well suited to the conditions in which they live. An example is sharks, which have barely changed in 300 million years. If a change happens in the environment, organisms must adapt, move, or die.

The pale and dark forms of the peppered moth.

A change can be slow, such as uplift producing mountains over millions of years, or very rapid, such as the arrival of a new predator or competitor in the area. The current increases in atmospheric carbon dioxide and temperature, pollution and deforestation all put evolutionary pressure on organisms. These are challenges which require rapid adaptation. They will produce a burst of evolution for some organisms, and extinction for many others.

QUICK CHANGE

Some organisms adapt very quickly. It's easiest for adaptations to come about rapidly in organisms with a short reproductive cycle, as they can pack more generations into a short time span. The peppered moth went from a predominantly white wing pattern to a predominantly black wing pattern during the Industrial Revolution when the trees the moths sheltered on became blackened by soot. White moths were readily spotted by predators against the black bark and were eaten. Darker moths were camouflaged and survived to breed. As soot in cities reduced, paler moths have returned. In China, the plant *Fritillaria delavayi* offers another recent example. The plant originally had yellow flowers, but as it has been picked for use in medicines, a form with grey flowers has become dominant. The grey flowers are hard to spot against the rocky landscape where it grows, so are more likely to survive than the conspicuous yellow flowers.

44

Evolution, now with genes

The modern evolutionary synthesis

When Darwin set out his theory of evolution by natural selection, he didn't know how inherited characteristics are passed between generations, or how changes appear in organisms. These were explained by the work on heredity and genetics in the 20th century.

Information from paleontology (the study of fossils), embryology, mathematics and other disciplines has also fed into evolutionary theory. The result was a reappraisal and restating of evolution in terms of genetics, commonly called the 'modern evolutionary synthesis'.

NATURAL SELECTION IS NOT THE ONLY WAY

Darwin explained evolution in terms of natural selection: when there are slight variations between organisms, those best suited to surviving in their environment will succeed, passing on their features to the next generation. In this way, the beneficial features are strengthened over time. Modern evolutionary theory recognizes additional ways in which evolution can take place. One of these is 'gene drift' which is random. Chance events cause one allele (form of a gene) to become more prevalent than others in a population, or cause a feature to die out as its allele becomes infrequent, just through the random chance of how organisms come together and breed. Gene drift is most common in small populations, where a slight shift in the prevalence of an allele can have a large impact. (An allele is the version of a gene that produces a particular characteristic. So a gene for fur colour might be available as an allele for brown fur and as an allele for black fur.)

Sexual selection works alongside natural selection, but can lead to seemingly inconvenient traits becoming common. An example is the tail of a peacock. It must be an encumbrance for a male peacock to have such a large and heavy tail, but it's been selected because peahens prefer and breed with males that have large tails. Perhaps the large tail indicates that the male can find enough food and keep territory in spite of his tail, so he must be a fit specimen. It's a bit like young men who court danger appealing to potential partners.

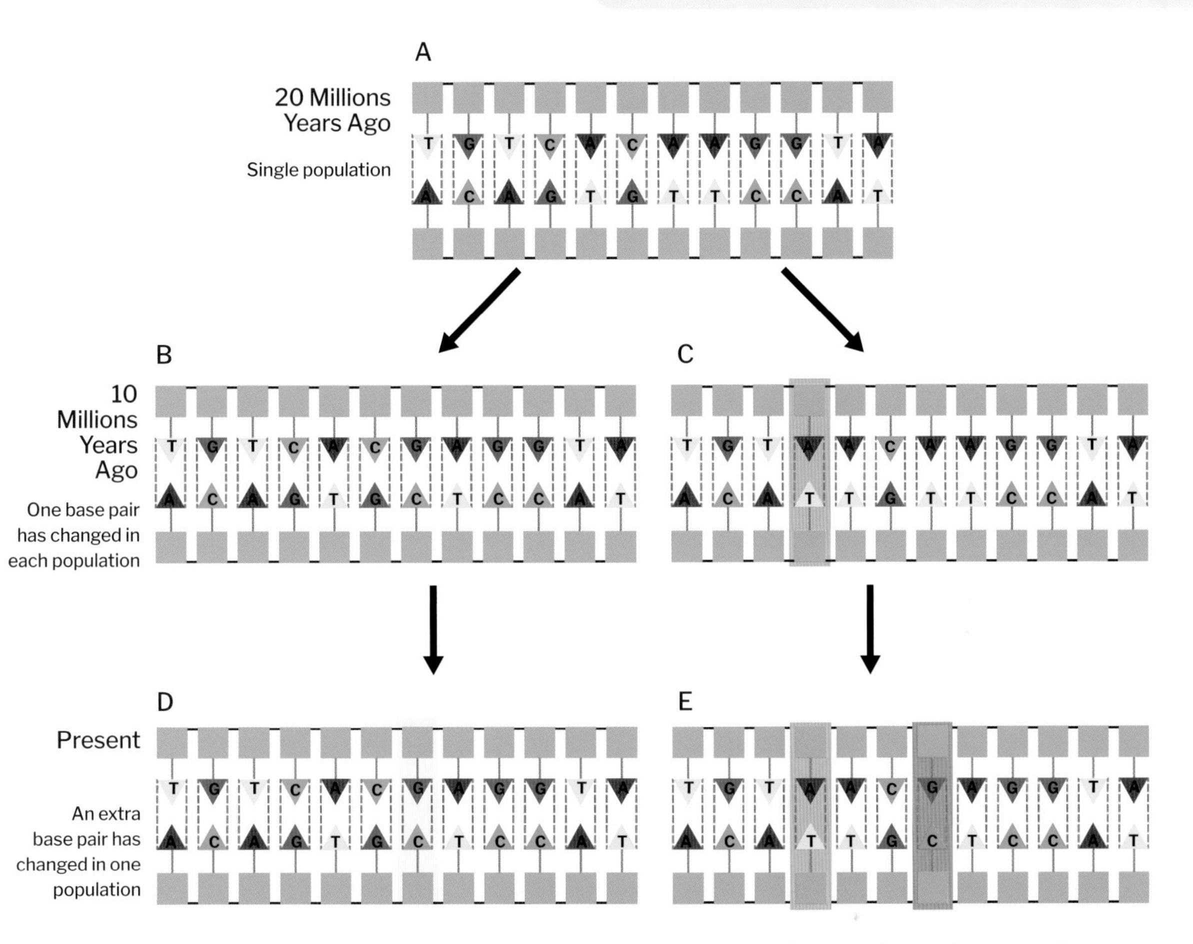

A molecular clock works backwards from the present by looking at the rate of mutation to work out when species diverged.

EVOLUTION AS ALLELES

In 1937, the Russian biologist Theodosius Dobzhansky wrote the first book that brought evolution and genetics together. He redefined evolution in terms of 'a change in the frequency of an allele within a gene pool'. As an allele becomes more and more common, the feature it produces becomes prevalent in the population. Dobzhansky pointed out that many different evolutionary changes can be happening in an organism at the same time, and that mutation drives evolution. A mutation is a random change, caused by a mistake in copying genes. It can produce harmful, harmless, or beneficial variations. If a beneficial or harmless characteristic is passed on, it can become the most common, changing the species.

LEAPS AND BOUNDS

Darwin envisaged evolution as a slow and steady process, but this assumption was overturned in the 20th century. Dobzhansky found that evolution could move quickly. A model of 'punctuated equilibrium' suggested that long periods of little or no change in an organism can be interrupted by short periods of rapid change brought about by a change in conditions. This can lead to speciation – a group of organisms changing to such a degree that they become a new species, distinct from their original community. Speciation occurs readily if a population is isolated. It can't reproduce with its original group and evolves from the restricted gene pool (available genetic variety) of the small group.

FROM NATURAL HISTORY TO BIOCHEMISTRY

The new emphasis on the mechanism of inheritance explained evolution in terms of genes and alleles, taking account of the location of genes on a chromosome. When chromosomes are separated in meiosis to produce egg and sperm cells, they undergo 'recombination' when chunks switch between corresponding chromosomes. This increases variety in future generations. Instead of offspring inheriting a whole chromosome from a parent, it can inherit different genes on the same chromosome from each parent. When chunks switch around, genes that are close together on the chromosome are more likely to move together, and so be inherited together. This kind of molecular-level insight into the process of inheritance cast a new light on evolution. Studying the rate at which a gene changes, in terms of changed base pairs over long periods of time, can be used to work out roughly when species diverged. This method of calculating speciation backwards through time is called a 'molecular clock'. It has only become possible since DNA sequencing became readily available.

Theodosius Dobzhansky.

45

Dead and gone

Extinction

Today, it seems entirely obvious that some organisms (in fact, most organisms) have gone extinct in the past, in the process of evolution producing the current inhabitants of the planet. Their fossils are all around us, and sometimes preserve traces of bodies that are nothing like any living organism. But just 250 years ago, scientists were still arguing whether extinction ever happened.

CUVIER AND THE LOST ELEPHANTS

The French biologist Georges Cuvier studied the fossil remains of elephants found near Paris, where elephants no longer live, and realized that they were substantially different from the bones of elephants then living in Africa. He found further differences between them and the remains of 'elephants' found in Siberia, which we now call mammoths. Cuvier proposed that they were a different type of elephant which no longer lived in the world. At the time, people struggled to accept the idea of extinction as it didn't fit with the teachings of the Christian Church: if God had created a perfect world, why would anything go extinct? Did it mean God had created imperfect animals that couldn't survive? Or had He got rid of some perfectly good animals?

Georges Cuvier.

It didn't seem to make sense. Instead, when fossils didn't look quite like living species, people tried to argue it away, suggesting that these variants still lived on somewhere remote and just hadn't been found yet. It would be hard to hide something as large as an elephant, though.

Cuvier went beyond saying some animals go extinct, to proposing periods of 'catastrophe' when very many species face extinction at the same time. He thought that the development of Earth and life on Earth was punctuated by these periods of immense change. It continued steadily for a long time between disasters. Cuvier proposed in 1813 that these disasters were often floods.

MASS EXTINCTION IS A CATASTROPHE

Cuvier's model, catastrophism, became quickly unpopular after other scientists tried to link it to the flood of Noah described in the Old Testament – a connection Cuvier never suggested. The geologist Charles Lyell popularized the view that the geological processes that shape Earth are very slow and ongoing. Charles Darwin, proposing his theory of evolution in 1859, outlined a process of very slow and steady change in living organisms without catastrophic events.

Today, scientists recognize two patterns of extinction: continual 'background' levels of extinction as organisms struggle to adapt to change and some fail, and occasional mass extinction events precipitated by catastrophic and often sudden change. 'Sudden', in geological terms, can still mean hundreds of thousands of years. There have been five mass extinction events since the evolution of complex life forms, and several less severe extinction events between them. The severity of an extinction event can be measured in several ways: the number of families of organisms that go extinct, the number of genera, number of species or estimated number of individual organisms. Mass extinction events are those in which more than 75 per cent of species die in a relatively short period – less than 2.8 million years.

EXTINCTIONS, PAST AND PRESENT

Mass extinction events have had several causes. The most famous occurred 65.5 million years ago at the end of the Cretaceous period, and wiped out all the non-bird dinosaurs, pterosaurs and ammonites, among others. It was unusual in apparently being caused by a single, sudden event: an asteroid crashing into Earth in what is now the Gulf of Mexico. The immediate effects would have been catastrophic tsunami in the local area, worldwide fires, volcanic and earthquake activity, followed by drastically falling temperature and light levels as the atmosphere filled with dust, blocking sunlight.

Climate change is a deadly feature of mass extinction events, though the causes

The Chicxulub impact event which caused the extinction of the dinosaurs.

differ. While an asteroid caused cooling that killed the dinosaurs, the most deadly of all mass extinction events resulted from climate chaos caused by up to a million years of devastating volcanic eruptions. The end-Permian extinction, 252 million years ago, is called the 'Great Dying'. It killed around 95 per cent of all species on Earth, a disaster from which the planet took millions of years to recover. Some other extinction events have probably been caused by living organisms. Mosses

Bring in the new

Extinction events clear out a lot of habitats and niches where organisms once lived. These are then available to new inhabitants. Organisms move in, and adapt to make the most of their new situation, giving rise to new species. This is how biodiversity is eventually restored after mass extinction events.

and algae growing on land took carbon dioxide from the atmosphere when they evolved and spread in the Ordovician period, causing a drop in temperature that produced a devastating glacial period (ice age) 35 million years ago. It's been suggested that organisms were similarly responsible for a series of global glaciations – Snowball Earth events – starting around 720 million years ago. And cyanobacteria likely prompted the original Snowball Earth event 2.3 billion years ago by removing carbon dioxide from the atmosphere.

SLOW AND STEADY

Alongside the occasional extinction events, organisms do become extinct at a steady, slow pace of about one in every 100,000 species every century. As environmental conditions change, organisms must adapt to new conditions or move to a more favourable location. If they can't do that, they are likely to die out. Sometimes they will be driven to extinction by the emergence of a competitor or predator – either newly evolved or moving into their area. At other times it might be environmental change, such as a river changing its course, or climate change, that pushes organisms over the brink.

PICKING UP THE PACE

Humans have already overseen the extinction of many species. As early hunters, we probably played a part in the destruction of very large animals (megafauna) in most of the world. Through habitat destruction, hunting and changing environments, we have driven many more, and many less readily noticeable, species to extinction. Many scientists believe that the sixth mass extinction is already underway and that we have caused it.

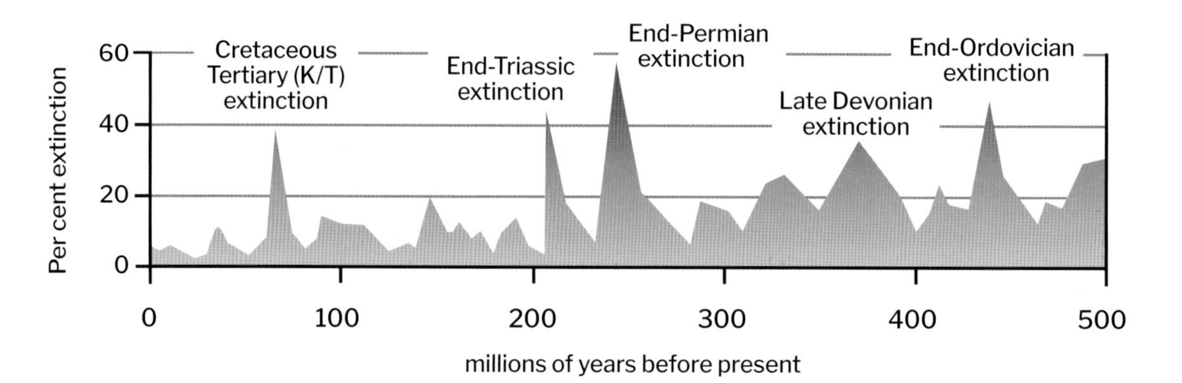

Levels of extinction over the last 500 million years.

46

It's all relative

Special and general theories of relativity

Albert Einstein is possibly the most familiar name in science, and many people can recite his famous equation, $E=mc^2$, even if they're not sure exactly what it means. The equation is part of Einstein's special theory of relativity, published in 1905. He produced a fuller theory, general relativity, in 1915. Einstein's work was originally theoretical, but it has since been supported by evidence in the form of observations and experiments.

SPECIAL RELATIVITY

The special theory of relativity defines the relationship between speed, time, mass and space. It relies on the assertion that the speed of light is constant in a vacuum and is the same for all observers, regardless of whether they are moving. Speed affects time, mass and space – at least, very high speeds do. At speeds approaching the speed of light, mass increases and time (as experienced while moving) goes more slowly.

Einstein began with a thought experiment. Imagine a train moving at nearly the speed of light. When the train is exactly halfway between two trees, both are struck by lightning at the same time. An observer beside the track experiences both lightning strikes simultaneously. But a person on the train would see lightning strike the tree ahead first and then the tree behind. They are moving towards the tree ahead, so light has less distance to travel to meet the observer. They are moving away from the tree behind, so it has further to travel and takes longer to meet the observer. From this, we can see that time is relative and depends on the 'frame of reference' of the observer – the place in space and time from which an event is viewed. The theory is 'special' in that it relates only to frames of reference with no acceleration.

Time moves more slowly when an object travels near the speed of light. So a person in a spaceship travelling at almost the speed of light would experience less time passing than a person standing on Earth for the same duration. If you could travel at almost the speed of light for five years (as experienced by you) in

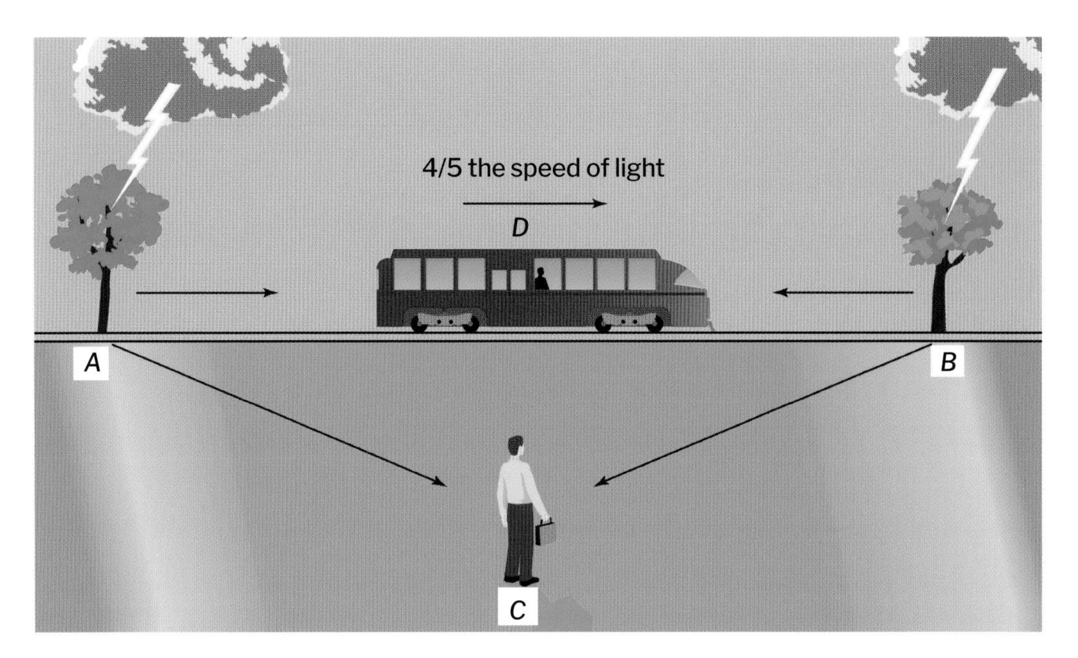

Light from A and B reaches observer C at the same time, but for observer D, the lightning strike at B seems to occur first.

a spacecraft, a person on Earth would have to wait 50 years for your return. Atomic clocks on satellites have to 'lose' seven microseconds a day to keep in synch with atomic clocks on Earth, even though satellites don't travel anywhere near the speed of light.

The mass of an object increases as it speeds up, until the mass of an object travelling at the speed of light would be infinite. For this reason, faster-than-light travel is deemed to be impossible.

$E=mc^2$

Einstein's famous equation tells us that matter and energy are interchangeable. A tiny amount of matter can be converted into a massive amount of energy. In the equation, *m* is mass and *c* is the speed of light; *E* is the energy equivalent. If a paperclip could be converted just to energy, it would release as much energy as the atomic bomb that destroyed the Japanese city of Nagasaki in World War II. The equation – or the situation it describes – is what makes possible atomic power and weapons, using nuclear fission to produce energy from matter.

GENERAL RELATIVITY

Einstein went on to develop his theory to take account of gravity, missing from his considerations in the special theory. His theory of general relativity shows the fabric of the universe, made of spacetime, to be distorted by mass. Spacetime bends around a massive object (that is, an object with mass), and this curvature produces the effects of gravity. In an analogy in two dimensions, it's often compared with holding a blanket taut and then dropping a heavy ball on to it. The ball causes a dip in the blanket, and any other objects will roll towards it – not because they are attracted to the mass of the large ball, as Newton would have explained it, but because the curvature of the surface changes their path. It's a bit harder to visualize in the four dimensions of spacetime, but the effect is the same: the distortion of the spacetime continuum means things fall towards a massive object. The more mass the object has, the more spacetime is distorted and so the more attractive the object appears to be.

Einstein's model of gravity laid out in this theory predicts that even light should be bent by gravity – not because it has mass but because the path it takes goes through a curved medium. If you followed a straight path between your house and a distant building, you would actually travel in a curve because the surface of Earth is curved. Similarly, light going 'straight' through space might be constrained to follow a bent path because spacetime is itself bent.

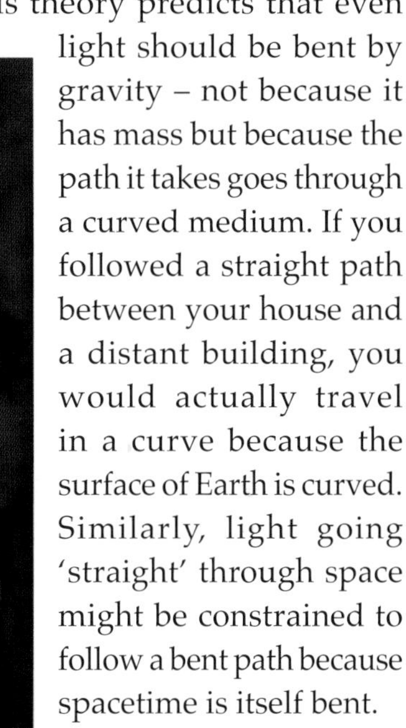

Albert Einstein.

47

Smaller than small

Quantum physics

Quantum physics is the physics of the very, very tiny – at the subatomic scale. It's pretty weird and wonderful stuff. It explains how atoms work, so how all chemistry and biology are possible. Much of the modern world, from MRI scanners to the Internet and from lasers to nuclear power stations, relies on quantum physics and exploits quantum phenomena.

THROW AWAY THE PARTICLES

The way we generally learn physics at school, and even talk about reality in everyday life, is with the universe made up of particles that have actual physical presence. We draw electrons, neutrons and protons as little circles and say that all matter is made of them (as we have done elsewhere in this book). Quantum physics unpicks all that, which makes it alarming. Instead of thinking of particles as discrete blobs with a defined presence in space, quantum physics sees them as little waves, bursts of energy of a fixed size (called quanta).

Physicists first realized in the 1920s that the wave-particle duality of light also applies to other particles. Electrons fired through two slits in a re-run of Thomas Young's double-slit experiment with light (see page 80) also produce interference patterns. That is, electrons work like waves too. Instead of drawing an electron as a circle going around a nucleus made up of other circles (neutrons and protons), quantum physics shows an electron as a wave function, like this:

WHERE IS IT?

Unlike a little blob of matter, a wave function doesn't occupy a defined position in space. To say where an electron is – or, rather, is likely to be – physicists square the amplitude (height of the wave) and call this the 'probability distribution'. It gives an area in which the electron is most likely to be at any time. In reality, it could be anywhere, but it's probably within this space.

This randomness as a feature of the universe is unnerving, and is quite the opposite of the entirely predictable Newtonian physics in which everything follows certain rules. That seems to work with objects at a large scale, but not at a very small (quantum) scale.

We can't be sure the wave function model is correct, as whenever an electron is measured, whenever its location is discovered, it appears as a particle. When we look at a subatomic particle, it's no longer a waveform. The act of measuring quantum events changes them, fixing their uncertainty into a form of certainty (see page 194). The movement from probability (wave function) to fixity is called 'collapsing the wave function', but we can't explain how it happens.

ALL THE WEIRDNESS

Some aspects of quantum behaviour look so odd they are hard to grasp. At a quantum level, particles seem to be capable of being in two places at once, can spin in all directions simultaneously, can wink in and out of existence and can respond instantaneously, apparently with faster-than-light communication. The weirdness partly comes from the fact that we are constrained in our explanations by the language we use. A large object can't spin in two directions at once: if we say a particle can, that really means 'spin' or 'direction' has a slightly different meaning. A particle can be 'in two places at once' if it is seen as a wave function. This has been demonstrated repeatedly using, again, the double-slit experiment. If photons (or electrons, atoms or even molecules) are fired through the slits, they form interference patterns indicative of waves. And they do this *even if they are fired one at a time* towards the slits. The interference patterns are made by waves going through both slits interfering with each other – reinforcing in some places and cancelling in others. Yet if photons are fired one at a time, they have nothing to interfere with except themselves. They can only make interference patterns if they each go through both slits at the same time. Don't even ask.

The so-called 'Copenhagen interpretation' of much quantum weirdness is that nothing becomes fixed until we observe or measure it. Einstein dismissed this, saying that he liked to suppose the Moon was there whether he looked at it or not. The alternative, the 'many worlds' explanation of quantum weirdness, is more bizarre. In this explanation, all the myriad possibilities are all true all the time. So every quantum particle is in every state it might be, all in parallel universes inaccessible to us. This scales up to real-world parallel universes. You are reading this book in one universe, but are perhaps climbing a mountain in another.

PARTICLES AND FORCES

Physicists recognize four fundamental

Quantum tunnelling brings us sunlight

The Sun and other stars are powered by the fusion of hydrogen into helium. This requires protons (hydrogen nuclei) to come together and fuse – but protons are both positively charged, so they should repel each other and not come together at all. The immense pressure at the heart of the Sun, produced by the Sun's own gravity, brings them into close proximity, but the reason they actually fuse is a phenomenon called 'quantum tunnelling'. Once we think of the protons as wave functions, they have a small probability of existing one inside another – of tunnelling through the barrier that separates them, and so combining. The process releases energy, in the form of photons which then slowly (and randomly) make their way out of the Sun.

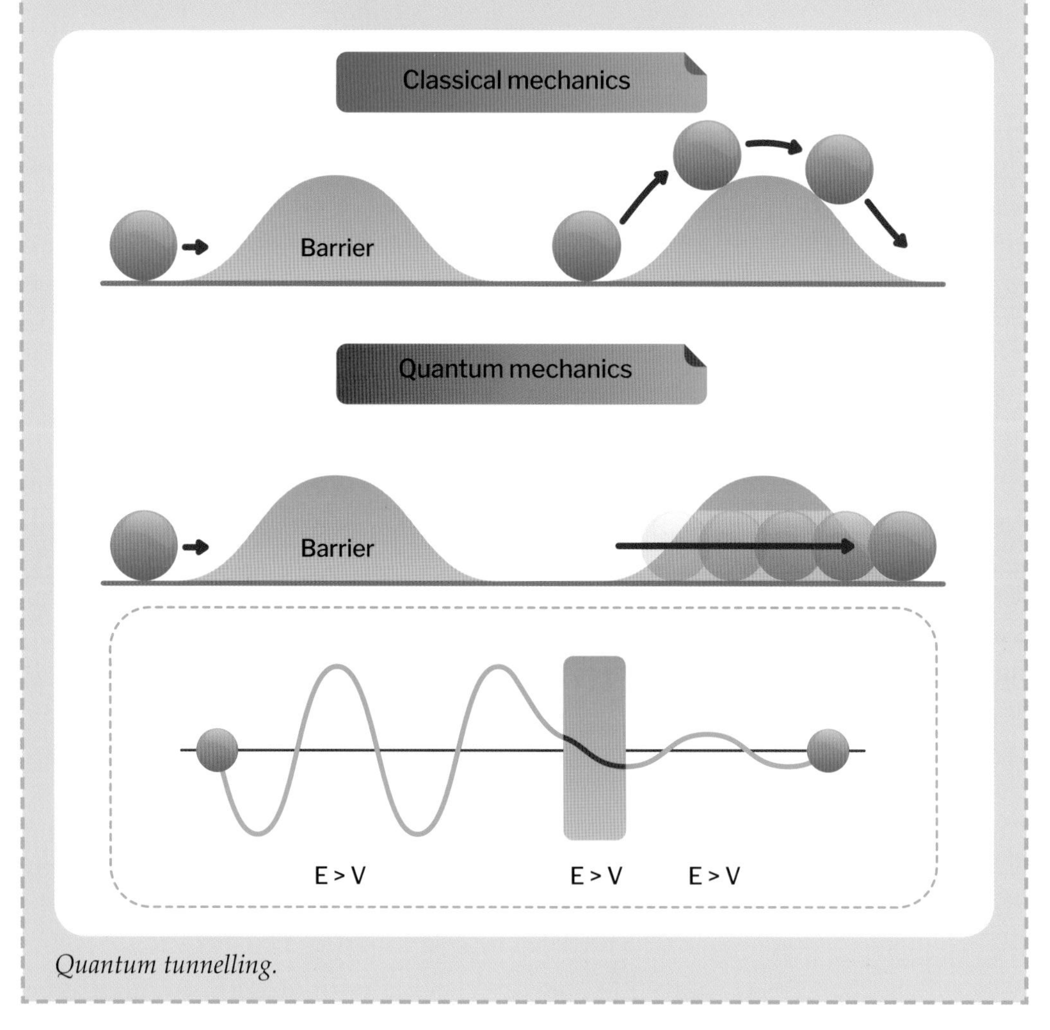

Quantum tunnelling.

forces at work in the universe: the weak nuclear force, strong nuclear force, electromagnetic force and gravity. Quantum physics can deal well enough with the first three, but so far gravity has eluded it.

The strong nuclear force, mediated by gluons, operates within the nucleus, holding the protons and neutrons together. Protons, both having a positive charge, naturally repel each other; the strong force overcomes this, holding nuclei together and so enabling atoms to exist. The weak nuclear force enables radioactive beta decay (see page 61). The electromagnetic force works to hold atoms and molecules together, with the negatively charged electrons attracted to positively charged protons. Gravity works on a much larger scale, holding stars, solar systems and galaxies together. It operates between objects that have mass.

The principal problem with quantum physics, apart from its weirdness, is that it can't accommodate gravity. Some physicists think that this means our theory of gravity needs rethinking and is incomplete. Quantum physicists have long suspected that the mass of quantum particles is so tiny, gravity has so little impact on them that it can be discounted. Mathematician Roger Penrose suggests that gravity is too weak to have an impact at the quantum level, and so 'allows' particles to be in two places at once. But this takes energy to maintain, so can't be sustained at larger scales. Gravity forces larger objects to snap into one place or another.

In quantum field theory, forces are mediated by force-carrying particles called bosons which interact with fermions (the particles that produce matter, which are quarks and leptons). The strong nuclear force holds the bits of the nucleus together, and also, paradoxically, keep them sufficiently far apart that the nucleus can exist. (The strong force becomes repulsive at very close quarters, though attractive at slightly larger distances.) The gluons hold together the quarks that form protons and neutrons (see page 128), and less directly hold the neutrons and protons in place in the nucleus. The weak nuclear force is mediated by three types of bosons that exist briefly during the beta decay of radioactive atoms. The electromagnetic force is mediated by photons. Gravity is possibly mediated by gravitons, but their existence has not been confirmed.

'SPOOKY ACTION AT A DISTANCE'

One of the strangest aspects of quantum physics is entanglement. If two particles are entangled, their states are linked. Suppose the particles have opposite states: one has clockwise spin and the other has counter-clockwise spin. If they are then widely separated, their entanglement remains. If one is measured, the state of the other is instantly known, as the relationship between their states is the same. If the local particle has clockwise

String theories

String theory is a development of quantum physics that tries to see particles as different types of vibration of tiny looped 'strings'. Just as a string on a musical instrument makes a different sound if it vibrates differently, the strings of string theory produce different particles according to how they vibrate. String theory has proved mathematically useful, and can even incorporate gravity, but it requires the universe to have ten dimensions for it to work. Trying to make it fit our universe, with three dimensions of space and a fourth of time, has so far failed. It might prove to be a useful tool but probably not, as people once hoped, a unifying theory that explains everything.

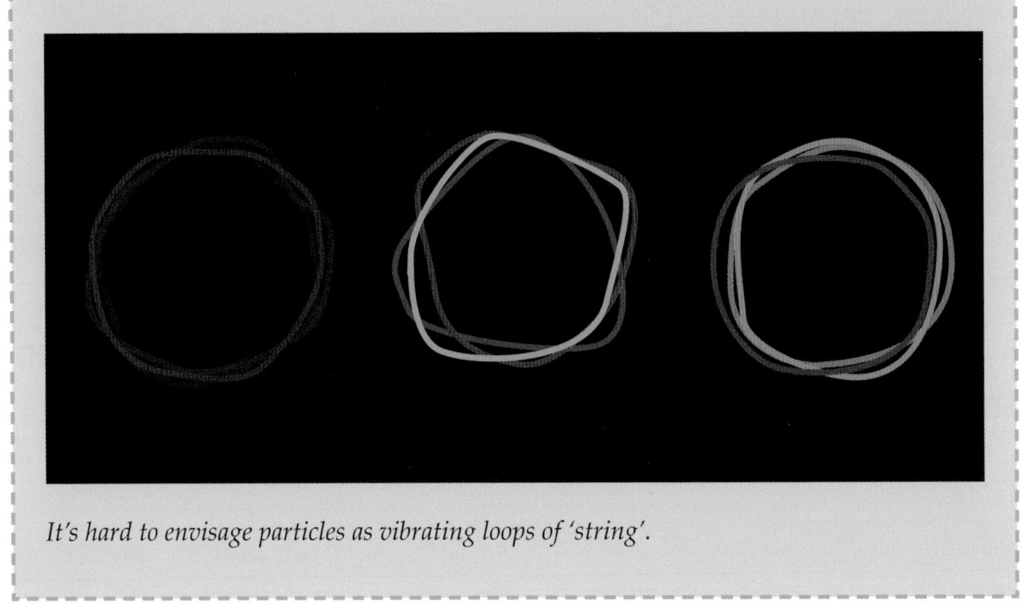

It's hard to envisage particles as vibrating loops of 'string'.

spin, we instantly know the other particle has counter-clockwise spin at the same moment, even if it's the opposite end of the universe. As measuring the spin collapses the wave function and crystallizes the particle's state, measuring one has an impact on the other. Einstein called this 'spooky action at a distance'. He wasn't a fan.

48

A cat in a box

The uncertainty principle

The 'uncertainty principle', derived by physicist Werner Heisenberg in 1927, is fundamental to quantum physics. It explains how, at a subatomic level, we can't know everything about a particle: the very act of measurement changes what is being measured. This is not a limitation of our ability to measure particles, but is a feature of the way particles behave. It's a fundamental and unavoidable feature of the universe as quantum physics describes it.

The position of a particle considered as a wave function (see page 189) is given by the amplitude (height) of the wave squared, which gives us the probability distribution. We don't know the position of the particle exactly – the probability distribution is a guide as to where the particle is most likely to be found.

The momentum of the particle is given by the length of the wave. The momentum is also not known with precision as the wave function isn't a single, uniform wave; it has many different wavelengths.

amplitude

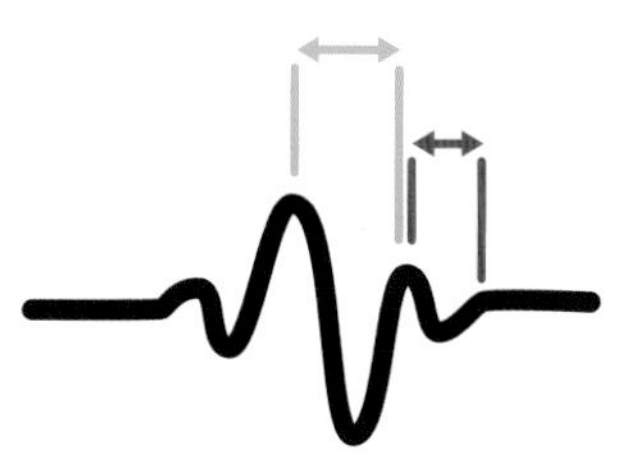

many different wavelengths

Probability distribution gained by squaring the amplitude

The bottom of these two shows the probability distribution gained by squaring the amplitude of the wave function at the top.

If we pick a wave function with a single wavelength (so it has known momentum), squaring the amplitude to give the probability distribution gives a very unhelpful curve:

probability distribution

The probability distribution goes on like this infinitely. The particle is equally likely to be found anywhere in the universe. By picking a single value for momentum, we lose the ability to find the location of the particle. It works the other way around, too. If we know the exact location of a particle, we can't tell its momentum.

DON'T LOOK NOW

The very act of observation changes what is being observed. Although photons or electrons fired through a double slit, even one at a time, produce interference patterns, if a detector is installed on one of the slits, the beam starts to behave like particles.

The most famous explanation of this is the thought experiment known as Schrödinger's cat. Originally intended to show how ridiculous the concept of this aspect of quantum physics looks, this has now become famous in its own right as a way of explaining the quantum effect of superposition (things being in two states or places at once).

Imagine a cat in a box. Sharing the box with the cat is a source of radioactivity, a detector and a vial of poison. Radioactive decay is unpredictable. We know the probability that a particle will undergo decay in a particular interval, but can't be certain that it will. In the thought experiment, the decay of a radioactive particle triggers the release of the poison in the vial which immediately kills the cat. The death of the cat is, then, related to the probability of decay. The idea is that until measured (observed) the radioactive particle is in both states, decayed and undecayed. Consequently, is the cat in both states, dead and alive? As soon as we observe the cat, the wave function collapses and the cat is either one or the other, just as the particle is either decayed or undecayed.

Shrödinger's cat, dead and alive.

49

Gaia

The living Earth

The idea that Earth is something like a living organism that maintains the right conditions for life (particularly human life) is very old in spiritual traditions. In the 1970s, the English chemist James Lovelock proposed it as a scientific model. Gaia theory describes Earth as a self-regulating system that balances itself in response to stresses, maintaining conditions favourable to life.

FINDING A BALANCE

It's well established that living things interact with the planet's geology, atmosphere and water in complex ways. Our atmosphere contains enough oxygen to support current life, but oxygen must be constantly renewed by the activity of plants and photosynthesizing bacteria. Carbon constantly cycles through the atmosphere, living things and rocks.

Intermittently, there are massive assaults on the equilibrium. But over time (often millions of years), the system rebalances and reaches a new equilibrium. For example, around 645 million years ago, the climate cooled dramatically, probably as organisms removed carbon. Most of Earth was covered with ice. The cold reduced the biological activity that removes carbon dioxide, but it didn't reduce the volcanic activity that returns carbon dioxide to the atmosphere. Over millions of years, the atmosphere regained enough carbon dioxide to warm the world and life resumed, building up again.

FEEDBACK LOOPS

Gaia supposedly works through a series of feedback loops to favour life. If one condition changes, it impacts other parts of the system, ultimately looping into a corrective shift. In Earth's early history, the Sun produced 30 per cent less energy, and Earth should have been too cold for liquid water. Lovelock proposed that early micro-organisms produced methane, a powerful greenhouse gas. Methane trapped enough heat to maintain a habitable temperature, so nurturing life. That the earliest microbes were probably methanogens is now widely accepted.

ARGUING FOR LIFE

Where Gaia theory differs from some other formulations of connected sys-

Daisyworld

In a computer simulation known as Daisyworld, Lovelock and Andrew Watson showed how competing organisms can produce stable conditions. In a theoretical world inhabited only by daisies, white daisies reflect light and heat, cooling the planet, while black daisies absorb light and heat, warming the planet. If too many of one colour grows, the planet becomes too warm or too cold. Populations naturally rebalance to keep temperatures within a survivable level. Adding further species to the model gave a surprising result: it became more resilient – a precious lesson in the value of biodiversity.

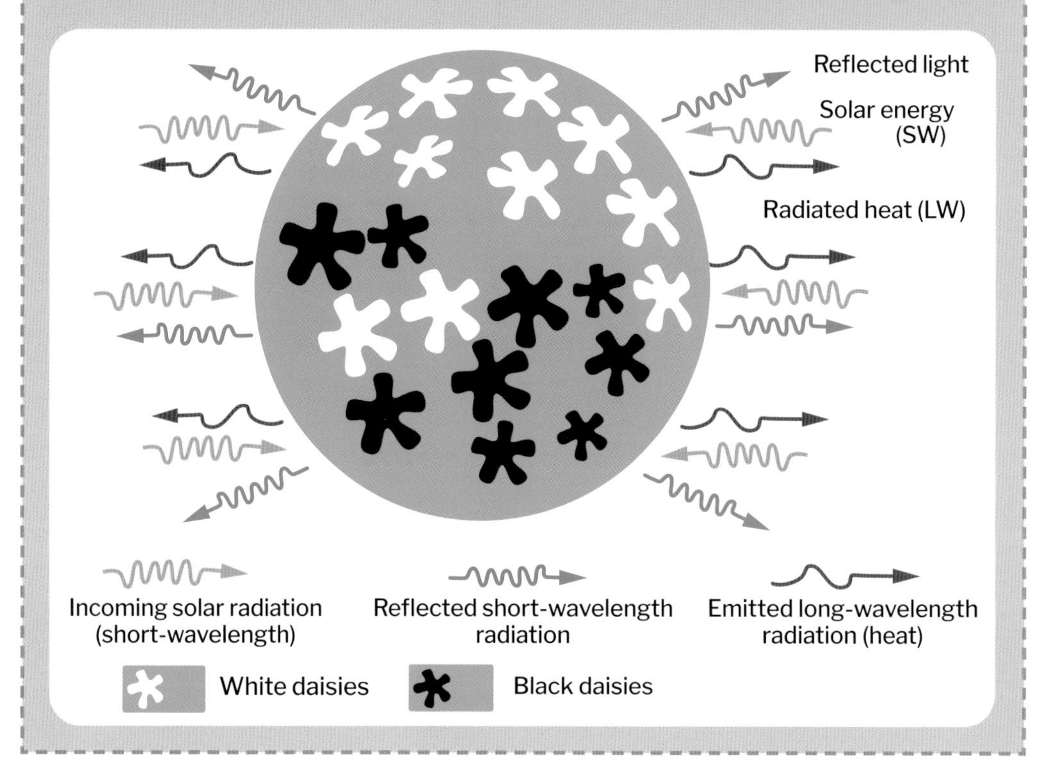

tems is in proposing that it actively promotes and protects life. This led to considerable criticism from many other scientists. There are now statements of 'weak' and 'strong' Gaia theory. Weak Gaia theory states the non-contentious view that there is a connection between living and non-living systems. Evolution affects and is affected by the environment – so living things change the atmosphere or surface temperature, for instance. Strong Gaia theory has Earth itself working like a giant organism, striving to maintain conditions. There is no scientific evidence or test for this idea.

50

Are we alone in the universe?

Extraterrestrial life

In the popular imagination, extraterrestrial life (or ETs) are green aliens in flying saucers or similar craft. In reality, extraterrestrial life is anything living that has its home outside Earth. It could be an advanced civilization on a planet in a different galaxy, à la *Star Trek*, or a colony of primitive microbes on an asteroid. Statistically, it's much more likely to be the second of these.

OTHER WORLDS

Assuming extraterrestrial lifeforms need somewhere to live, you can't have alien life unless you have other habitable worlds. Some spiritual and religious cosmologies have postulated other inhabited worlds, but as a scientific proposition the idea is relatively recent. The Italian philosopher Giordano Bruno wrote of 'innumerable worlds' with living beings in 1584 (after Copernicus proposed a Sun-centred solar system but before the invention of the telescope). He was burned at the stake for heresy in 1600. The belief in other worlds had been an established heresy since AD 384, showing at least some people were thinking about the idea. In 1609, Galileo Galilei turned his telescope to the planets and saw them resolved into disks. At that point, the possibility of other worlds became very real.

Bruno went beyond Galileo and the solar system, proposing that each star had its own solar system with worlds that could support living things as impressive as those on Earth. He was far ahead of his time: modern astronomers confirmed the first exoplanet (a planet outside our solar system) in 1992, and announced as recently as 2015 that most stars have planets in their so-called habitable zone (where life is theoretically possible). All that remains to be done is to find life beyond Earth.

LOOKING FOR LIFE

The easiest and most obvious place to search for life beyond Earth was within our own solar system. The Moon landings and robotic exploration of other planets has ruled out some places. In the 19th century, the idea of life on Mars became popular following the mistaken claim of some astronomers that they'd seen artificial waterways on Mars. As the planet most like Earth, it's remained the best local planetary candidate to support life. The current Mars Exploration Program by NASA is looking for evidence of past or present microbial life. There's no expectation that large aliens with a developed civilization will be found.

Planets aren't the only places that life might be found, though. Some of the moons of other planets, particularly Saturn and Jupiter, might be more promising places to look. A moon such as Saturn's Enceladus could potentially hide life. It has a frozen surface, but kilometres below the ice lies an ocean of liquid water. It could be sterile, or teeming with life – we have no way of knowing yet.

We know from life on Earth that living things can endure some terrible conditions. Microbes known collectively as 'extremophiles' live in places that seem to us very hostile: deep beneath ice, inside rocks, or in the scalding, acidic waters around hydrothermal vents in the sea, for example. The first living things on Earth might have emerged and lived in just such conditions. Most scientists, though, think that liquid water is essential for any type of life that we would recognize. Planets, moons, or even asteroids that have the right conditions, somewhere, for liquid water to exist are potential homes for extremophile microbes.

Looking for life beyond Earth in the solar system has focused on close examination of samples (such as rocks on Mars) and on working out the conditions

The Mars exploration rover Perseverance.

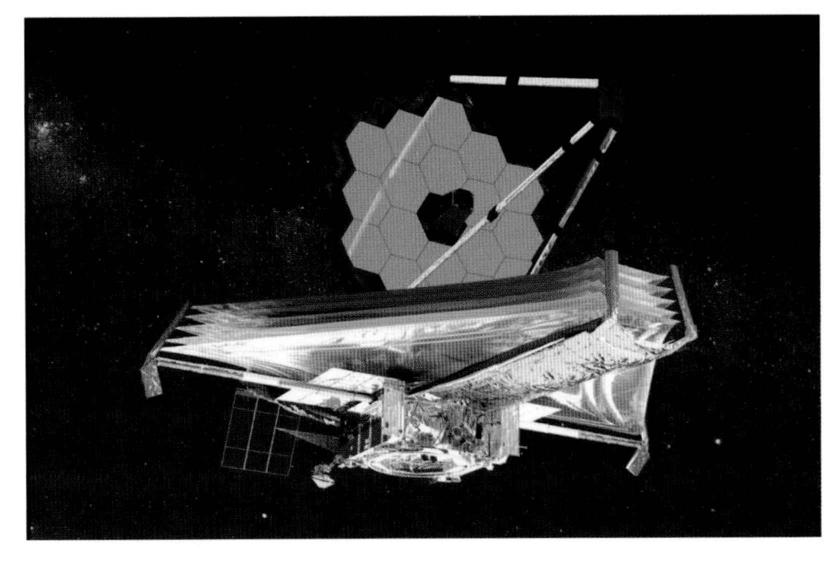

A 3D rendering of the James Webb Space Telescope.

and composition of surfaces, subsurfaces and atmospheres. Astronomers search for biosignatures – chemical signs that life is present – or just conditions that could be favourable to life.

SETI: THE SEARCH FOR EXTRATERRESTRIAL INTELLIGENCE

It's one thing to look for microbes in our local area of the solar system, and quite another to hope to find intelligent aliens. It looks pretty certain that there are no advanced alien civilizations in the solar system, but there could be some elsewhere. Indeed, given the statistics, it seems likely that there are. There are hundreds of billions of stars in our galaxy, and hundreds of billions of galaxies in the universe. If all or most stars have planets, the chances that none of the others supports life seem slim. How would we find them?

Using telescopes such as the new James Webb Space Telescope, launched in 2021, and before it the Kepler Space Telescope (2009), astronomers can hunt for exoplanets around distant stars. They can then work out whether a planet is close enough to its star to have liquid water. In 2011, Kepler found its first rocky exoplanet, but now hundreds are known. The James Webb Space Telescope will be able to explore the content of the atmospheres of planets and perhaps find the building blocks of life. These can show us where the potential for life exists beyond the solar system, but actually finding life is another matter. Within our solar system, we can send robotic craft to explore, but planets beyond the solar system are too far away for a visit. The closest exoplanets are 4.2 light years away. That means even a radio signal will take more than four years to reach the planet. Using the best existing space technology we have, it would take many thousands of years for a craft to visit.

If a spaceship we could build now had left Earth for Proxima Centauri at the time our *Homo sapiens* ancestors left Africa in large numbers 60,000 years ago, it would be only three-quarters of the way there by now.

The search for intelligence, rather than life of any type, must be carried out at a distance. Our best bet is to find radio signals or other types of electromagnetic radiation that seem to have an artificial origin. Radio telescopes search the sky for all kinds of radio signals. Natural objects (such as stars) produce electromagnetic radiation of many types, all across the spectrum. An intelligent being sending a radio transmission will use a narrow band. If their transmission is intended to convey information it will probably not be just regular pulses, as many natural objects produce, but will have complex patterns. Radio telescopes have been scanning the sky and collecting data for more than 50 years, but have so far found only one signal, nicknamed the Wow! signal, that looks out of the ordinary. It has still not been explained.

WHO'S SAYING WHAT?

If we ever find a radio signal from an alien civilization, it could be either a message deliberately sent out into space, or something unintentionally leaking out. Earth has been leaking radio and TV broadcasts on a large scale since the middle of the 20th century. If these were noticed by aliens, they could probably guess an intelligent community was behind them. Radio waves travel at the speed of light, so an alien 70 light years away could already be working to interpret a 1950s news broadcast. If the first alien radio signal we find is leaked unintentionally, it could be the alien equivalent of a *SpongeBob SquarePants* episode or an advert for a fizzy drink.

An alien civilization deliberately sending messages into space would probably try to make them decipherable, or at least recognizable. Earth has sent one radio message of this type. Called the Arecibo message (because it was sent from the Arecibo radio telescope), it decodes into a blocky image of a human, the telescope itself, a DNA helix and a few other bits and pieces. It's pretty hard for a human to interpret. The chances are that we would struggle to understand even a deliberate transmission,

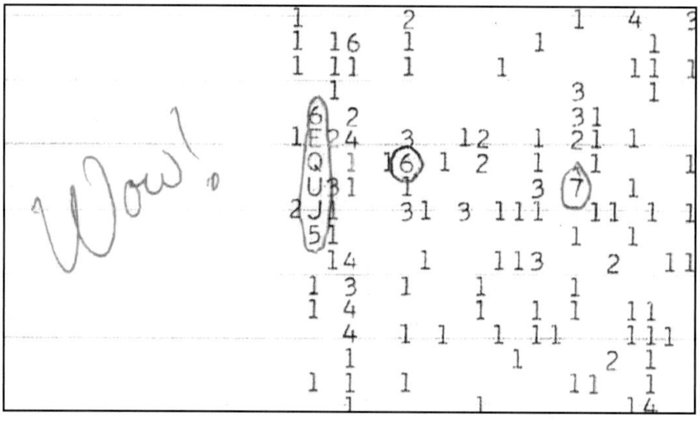

The Wow! signal, picked up in 1977 from the direction of Sagittarius.

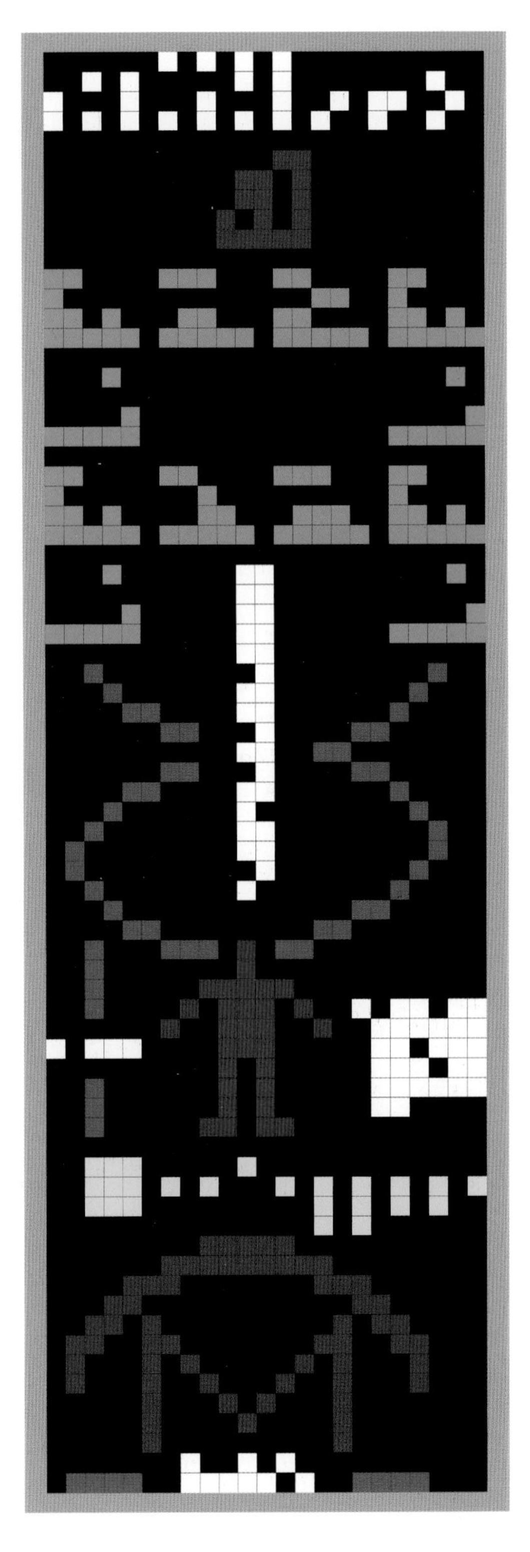

The Arecibo message (false colour) broadcast into space encoded in radio waves in 1974.

as aliens would probably struggle to understand ours.

Although we haven't sent more radio transmissions out into space for waiting alien eyes and ears, we have included messages on spacecraft. The two Voyager craft now heading out of the solar system into interstellar space both carry a 'golden record' of sounds and images from Earth. The Pioneer craft, launched in 1973, each have a plaque showing the position of Earth and images of a man and woman. Some scientists have questioned the wisdom of guiding aliens towards Earth: are we just inviting them to eat us, conquer us, or plunder our planet?

WHERE IS EVERYBODY?

Given the likelihood that there are billions if not trillions of planets that could support life, why haven't we found any aliens yet? Or why haven't they found us? That there should be many advanced civilizations out there, but none has been in touch, is called the Fermi Paradox, after the American-Italian scientist Enrico Fermi who posed the question, 'Where is everybody?' in 1950. A decade later, in 1961, Frank Drake wrote the so-called Drake equation to quantify the chances of our finding an intelligent civilization. It's a nice idea, but as we don't have values for most of the variables, it's not very informative:

Artist's impression of a Pioneer spacecraft heading into interstellar space.

$$N = R_* f_p n_e f_l f_i f_c L$$

It means that the number of technologically advanced civilizations in our galaxy, N, is equal to:

- R_* the rate at which stars form in the galaxy, multiplied by
- f_p the fraction of stars with planets, multiplied by
- n_e the number of planets per star with conditions that can support life, multiplied by
- f_l the fraction of suitable planets on which life evolves, multiplied by
- f_i the fraction of planets with life which have intelligent life, multiplied by
- f_c the fraction of planets with intelligent life that develop civilizations capable of transmitting signals we could detect, multiplied by
- L the length of time such civilizations actually transmit signals.

Estimates of the number of civilizations we could detect right now vary from less than one (effectively none) to more than a million.

There are many reasons why there may not be many, starting with the potential difficulty of getting life started on a planet at all. We just don't know how hard that is. It's possible life starts easily, given the right conditions, or Earth may be very rare in having life. We know it's taken more than four billion years for life on Earth to get to the stage where we can send and receive radio signals through space, but don't know how typical an interval that is, or whether some barrier usually prevents civilizations reaching this stage. We know that we have had radio for just over 100 years, but have no idea how long we will continue to transmit. What if a world war, killer virus, or catastrophic climate change wipes out civilization, ending our transmissions? Are there things that civilizations tend to do that lead to their destruction? Maybe having disastrous wars, developing destructive artificial intelligences, running out of resources, or falling prey to diseases. If each civilization only produces radio signals for 100–200 years, the chances of any overlapping and finding each other are not great.

Another possibility is that there are other radio-capable civilizations out there, and that they even know about us, but they are keeping quiet. Perhaps they have found it's dangerous to contact others, or perhaps they prefer to watch and wait.

WHAT IF?

It's not very likely we'll find evidence of intelligent life elsewhere in the near future, but the chances of finding, say, evidence of past microbes in the solar system are much better. What would it mean for us? It would have a profound philosophical impact. It might play havoc with religious beliefs, as previous revelations about the cosmos have done. How special are we if life of different kinds is found everywhere? If perhaps there are civilizations far more advanced than ours? Would we maybe treat other organisms on our own planet better if we felt less special? Although we can't tell what impact the discovery would have, it would surely be significant.

Index

abiogenesis 87–90
aether 86
al-Haytham, Ibn 79
Al-Kindi 79
al-Nafis, Ibn 54–5
anaesthetics 24–6
Anaxagoras 96
antibiotics 136–9
antisepsis 26, 132–5
Aristotle
 classifying organisms 143
 electromagnetic spectrum 86
 heliocentrism 50
scientific method 7
vacuum 13
atomic structure 19–23
atomic theory 9–12
Avenzoar 53
Avery, Oswald 124
Balick, Bruce 155
Beaumont, William 57
Becquerel, Antoine Henri
 electromagnetic spectrum 85–6
radioactivity 59
Beijerinck, Martinus 78
Bernard, Claude 58
Big Bang theory 42–5, 104
biogeochemical cycles 158–61
black holes 154–7
body modelling 53–8
Bohr, Niels 19, 21
Boltwood, Bertram 66
Bošković, Ruđer 11
Boveri, Theodor 18
Boyle, Robert 10
Bradley, James 81
Brown, Robert (astronomer) 155
Brown, Robert (botanist) 12
Brownian motion 12
Bruno, Giordano
 extraterrestrial life 197
 stars 96–7
Buchner, Eduard 163–4
Bunsen, Robert 97
cell theory 46–9
Chadwick, James
atomic structure 20
 nuclear fusion and fission 91
Chain, Ernst 136–7
chaos theory 170–1
Chargaff, Erwin 124
Châtelet, Emilie du 168, 169
chemical elements 27–30
China, Ancient
 anaesthetics 24–5
 chemical elements 27
 magnetism 120
classifying organisms 143–5
climate crisis 117–19
conservation of energy 168–9
conservation of matter 140–2
Copernicus, Nicolaus 47, 48
Crick, Francis 124
CRISPR-Cas9 127
Curie, Marie 60
Curie, Pierre 60
Curtis, Heber 42
Cuvier, Georges 181–2
Dalton, John
 atomic theory 11–12
 chemical elements 28
dark energy 99–104
dark matter 99–104
Darwin, Charles
 Earth accretion 63
 evolution 172–6, 178, 179
 extinction 182
 heredity 17
Davy, Humphry 25
De magnete (Gilbert) 122
Democritus
atomic theory 9–10
vacuum 13
Descartes, René 55
Descent of Man, The (Darwin) 175
DNA 17, 52, 88, 90, 124–7
Dobzhansky, Theodosius 178–80
Earth accretion 63–6
Earth structure 105–9
Einstein, Albert
 atomic theory 12
 black holes 154
 dark matter 100

general theory of relativity 186–7
gravity 39–40
light 82
quantum physics 192
special theory of relativity 185–6
electromagnetic spectrum 80, 83–6
Empedocles 27
endosymbiosis 51, 146–9
enzymes 162–4
Euclid 79
eukaryotic cells 48–9, 146–7
evolution 7–8, 172–80
extinction 181–4
extraterrestrial life 197–203
Fermi, Enrico 91
Fleming, Alexander 136
Florey, Howard 136–7
Franklin, Rosalind 124
Gaia theory 195–6
Galen 53
Galilei, Galileo
body modelling 55
gravity 37–8
heliocentrism 51–2
light 81
stars 96
Galvani, Luigi 57
Gamow, George 97
Geiger, Hans 19
Gell-Mann, Murray
atomic structure 20
standard model 128
general theory of relativity 186–7
germ theory 75–8, 132
Gilbert, William 122
Gordon, Alexander 134
Gould, Stephen Jay 176
gravity 36–41, 191
Greece, Ancient
antisepsis 132
atomic theory 9–10
body modelling 53
chemical elements 27
classifying organisms 143
Earth accretion 63
electromagnetic spectrum 86
heliocentrism 50
light 79, 81
magnetism 120
stars 95–6
vacuum 13–14
greenhouse effect 114–16
Guericke, Otto von 13
Hahn, Otto 91, 92
Halley, Edmund 109
Harvey, William 54–5, 56
Heisenberg, Werner 194
heliocentrism 50–2
Helmont, Baptista van 72
heredity 15–18
Hero of Alexander 79
Herschel, William 85
Hippocrates 132
Holmes, Oliver Wendell 134
Hooke, Robert
cell theory 46–7
microbes 71–2
Hoyle, Fred 44
Hua Tuo 24
Hubble, Edwin 42–3
Huygens, Christiaan 80
India, Ancient
anaesthetics 25
magnetism 120
Islamic world
antisepsis 132
body modelling 54
chemical elements 27
light 79
Ivanovsky, Dmitry 78
Jenner, Edward 150–1
Jīvaka 25
Kepler, Johannes 52
Kircher, Athanasius 108–9
Kirchhoff, Gustav 97
Koch, Robert 76–8
Lavoisier, Antoine 28, 29
Leclerc, Georges-Louis 64
Lehmann, Inge 108
Lemaître, Georges 43–4
Leopold, Prince 25
Leucippus 9
Leeuwenhoek, Antonie van 71, 72
light 79–82
Linnaeus, Carl 143–5
Lister, Joseph 134–5
Lorenz, Edward 170–1
Lovelock, James 195, 196
Lyell, Charles 63, 65, 175–6

maglev trains 123
magnetism 120–3
Marsden, Ernest 19
mass 39
Maxwell, James Clerk
electromagnetic spectrum 84–5
light 80
Meitner, Lise 91, 92
Mendel, Gregor 15–17
Mendeleev, Dmitri 31–2, 33–4
Messier, Charles 42
meteorites 69
Meyer, Lothar 32–4
Michelson, Albert 85
Michelson-Morley experiment 8, 85
microbes 71–4
Micrographia (Hooke) 46, 71
Miescher, Friedrich
DNA 124
heredity 17
Miller, Stanley 88–9
Mitchell, John 154
modern evolutionary synthesis 178–80
Moon 70
Morgan, Thomas 18
Morley, Edward 85
Morton, William 25
Needham, Joseph 73, 74
Newton, Isaac 38
atomic theory 10–11
body modelling 56
electromagnetic spectrum 86
gravity 36–7, 38
light 80
nuclear fusion and fission 91–4
Oppenheimer, J. Robert 154
Pacini, Filippo 76
Paré, Ambroise 132–3
Parmenides 13
Pasteur, Louis
enzymes 163
germ theory 78
microbes 74
Payne-Gaposchkin, Cecilia 97
Penrose, Roger 191
periodic table 20, 21, 23, 31–5
Perrier, Carlo 30
Perry, John 65
photoelectric effect 82
Planck, Max 80–2
planetary accretion 40, 67–70
Plato 86
Poincaré, Henri 170
Priestley, Joseph 25
Pringle, John 133
prokaryotic cells 48–9, 146–7
Ptolemy 46
quantum physics 188–94
quantum tunnelling 190
radioactivity 59–62
Réaumur, René de 56
Redi, Francesco
body modelling 56
microbes 73
Ritter, Johann 85
Römer, Ole 81
Röntgen, Wilhelm 85
Rubin, Vera 99
Rutherford, Ernest
atomic structure 19
Earth accretion 66
electromagnetic spectrum 85–6
radioactivity 60
Schleiden, Matthias 52
Schrödinger's cat experiment 194
Schwann, Theodor 52
scientific method 7–8
Secchi, Angelo 97
Segrè, Emilio 30
Semmelweis, Ignaz 134
Shapley, Harlow 42
Slipher, Vesto 43
Smith, William 167
Snow, John 76
Snyder, Hartland 154
Spallanzani, Lazzaro
body modelling 56–7
microbes 73–4
special theory of relativity 185–6
standard model 128–31, 192
Stanley, Wendell Meredith 78
stars 95–8
Steno, Nicolas
Earth accretion 63
superposition 165–6
Strassman, Fritz 91
string theory 192

Sturtevant, Alfred 18
subatomic particles 19–23
superposition 165–7
Sutton, Walter 17–18
Szilard, Leo 91
tectonic plates 110–13, 166
Teed, Cyrus Reed 109
Thales of Miletus 7, 95–6
Theophrastus 143
Thomson, J.J. 19
Thomson, William (Lord Kelvin) 64–6
Tyndall, John 114
uncertainty principle 193–4
Urey, Harold 88–9
vaccination 150–3
vacuum 13–14
Varro, Marcus 71
Vesalius, Andreas 53
Villard, Paul 60
Virchow, Rudolf 52
Victoria, Queen 26
Wallace, Alfred Russell 173
Watson, Andrew 196
Watson, James 124
Wegener, Alfred Lothar 110
weight 39
Wells, Horace 25
Wortley, Montagu 151
Young, Thomas 80
Zweig, George 128
Zwicky, Fritz 99

Picture credits

t = top, b = bottom, l = left, r = right

Alamy: 186

Caltech: 41b (JPL / R. Hurt)

CERN: 130b

David Woodroffe: 98, 139, 148, 167

ESA: 45t, 45b

ESO: 156, 157

Getty Images: 120, 138

Library of Congress: 65, 66, 153

NASA: 70, 83, 95, 101, 104, 155, 198, 199, 202

Science Photo Library: 56, 87, 108, 109, 116, 137r, 144t, 161, 180, 183

Shutterstock: 8, 16, 20, 27, 30, 35, 39, 61, 62, 73, 79, 80, 81, 85, 93, 102, 106, 111, 113t, 113b, 115, 119, 122, 123t, 123b, 125, 127, 129, 141, 142, 144b, 149t, 159, 160, 163, 190, 194

Wellcome Collection: 11, 12, 24, 25, 26, 54, 57, 58, 60, 74, 76b, 133, 135, 137l, 150, 175

Wikimedia Commons: 10, 14, 18, 29, 31, 32, 38, 43, 44b, 47, 48, 49t, 49b, 50, 51t, 51b, 52, 55, 59, 64, 68, 69, 71, 72, 76t, 77, 78, 82, 84, 89, 90, 92, 96, 103, 112, 130t, 131, 149b, 152, 166t, 166b, 169, 171, 172, 173, 176, 177, 181, 187, 200, 201